The Wonderful World of
Life in the
Sea

The Wonderful World of
Life in the
Sea

Andrew C. Campbell

Hamlyn
London · New York · Sydney · Toronto

Acknowledgements

Ardea: Hans and Judy Beste 49; R. J. C. Blewiit 29; D. W. Greenslade 32; Pat Morris 31, 45T, 82; Valerie Taylor 59, 60B, 67B, 71. Biofotos: 9, 10B, 11, 12T, 12B, 13, 14, 15T, 15B, 16T, 19T, 20, 21B, 24, 25, 26T, 27, 28T, 35B, 36/37, 39T, 42, 43, 48, 67T, 69B. Bruce Coleman Limited: Jane Burton endpapers, 23, 40/41, 60T, 62/63; 68; Jack Dermid 26B; Francisco Erize 81; Inigo Everson 88; Jeff Foott title page, Al Giddings 83; Giorgio Gualco 61; M. Timothy O'Keefe 51; Charlie Ott 19B; Oxford Scientific Films 17T, 85, 86T. Institute of Oceanographic Sciences: 86B, 87, 89T, 89B, 90B, 91, 92, 93T, 93B, 94. Natural History Photographic Agency: Anthony Bannister 17B, 56/57, 64B, 75T; J. M. Clayton 28B; Stephen Dalton 22; Brian Hawkes 10T, 21T; Harold Metcalf 70; James Tallon 84; Bill Wood 47, 65T. Oxford Scientific Films: 72, 73, 74T, 76, 78, 79. Photo Aquatics: 55, 66. Seaphot: Dick Clarke 80; Peter David 74B, 77, 90T; Walter Deas 39B, 52, 58, 65B; J. David George 30, 54, 75B; John Harvey 18; Richard Johnson 69T; Christian Petron 53B, 64T; Rod Salm 53T; Helmut Schuhmacher 35T. Kim Westerskov: 6/7, 33, 34, 38, 44, 45B, 46.

Published by The Hamlyn Publishing Group
Limited
London · New York · Sydney · Toronto
Astronaut House, Feltham, Middlesex, England
Copyright © The Hamlyn Publishing Group
Limited 1978
ISBN 0 600 38257 5

Phototypeset by Filmtype Services Limited, Scarborough, England
Colour separations by Culver Graphics, Slough, England
Printed in Spain by Mateu Cromo, Madrid

Contents

Introduction

Scientists believe that life, in the form of
simple single-celled organisms, originated
in the early seas. From such primitive
ancestors all present-day plants and
animals, both aquatic and terrestrial, are
thought to have descended.

The seas and oceans of the world cover seventy per cent of its surface. They separate the vast land masses of the continents and, added together, amount to an enormous volume of water. Everyone who has studied geography knows that there are different sea and ocean types. On the one hand, there are the shallow seas overlying the continental shelves surrounding the land which are strongly influenced by the geological and climatic conditions prevailing on nearby land masses. Great rivers dilute the sea water and deposit silt around the mouths of estuaries which may consequently always be changing their shape. These shallow seas, known as the neritic provinces, are important to man because they support fisheries and provide commercial waterways. On the other hand, there are the deep ocean provinces where conditions are little influenced by land. Ocean water tends to be remarkably constant in temperature and salinity at given depths.

The physical conditions of enclosed seas such as the Baltic and the Red Seas are controlled by the conditions prevailing on the land masses which surround them. In the case of the Red Sea, there is virtually no source of fresh water and, at the same time, a great deal of water is lost by evaporation. This causes a relatively high salt content in the sea's water for which it is well known. The Baltic Sea, at the opposite end of the scale has a high fresh water input from many rivers and a low evaporation rate, and is, therefore, brackish and more like a fresh water lake at its northernmost limits.

Climate and weather control the physical features of the shallow seas and so govern the distribution of living organisms inhabiting them. Plant and animal species in any community are subjected to pressures. These originate both from physical conditions, such as those already mentioned, and from biological forces – the availability of food for instance or competition for food or for living space, and success in avoiding predators. Marine plants and animals depend on water for almost everything. Oxygen is dissolved in it so they need gills; it is relatively dense and can support their bodies, so a strong skeleton is not so important; they may be buffeted by waves so that some physical protection has to be developed.

7

Food is obtainable not only in the form of plants and animals visible to the naked eye but also in microscopic form and as particles of decaying organic matter known as detritus. Such food may be gathered by what is known as filter-feeding, that is, by the pumping of water through a filter to trap the tiny organisms and detritus, which are then eaten. By allowing the detritus to settle and accumulate on the sea-bed as a deposit like dust on a carpet, it can then be 'hoovered' up. It can also be wrapped up in strands of mucus and then eaten.

Methods of reproduction reflect the conditions important for survival. It is not necessary to mate, if sperm and eggs can simply be shed into the water where fertilization can then occur.

Many filter-feeders are sedentary and hardly move, if at all. How do they disperse and colonize new places? They do so by having a free-swimming larval stage which develops from the fertilized egg and which swims around or floats free before it settles and turns into an adult. Terrestrial organisms run the risk of desiccation, due to sun and air. This is not a problem in the body of the sea, but it is on the shore.

Some marine animals maintain a correct level of salt in their bodies by means of complex physiological mechanisms. Others just allow themselves to become as salty as the surrounding sea water. Such forms, however, cannot cope with extreme fluctuations in salinity which may kill them.

The plant and animal kingdoms display a great diversity of forms and life styles. Life is thought to have originated in the early seas and it seems certain that from simple creatures floating in the 'primaeval soup' all present-day forms have evolved. Many types of animal have been identified by scientists and arranged into groups known as phyla. A phylum is a major division of the animal or plant kingdom containing all those species thought to have a common evolutionary origin. Today about thirty animal phyla are known and almost all have marine representatives.

Plants are not so diverse as animals. They are organized into six groups which some botanists also call phyla. The majority do not occur in the sea. The flowering plants are represented there by sea-grasses, which grow in sand, and mangroves, which grow in shore mud. Algae flourish in shallow water where they are represented by the familiar sea-weeds. Sea-weeds provide food and shelter for shore animals. Algae also occur as minute drifting organisms (phytoplankton) in the surface waters of the oceans. The word plankton means drifting life.

In the sea, as on dry land, plants are the originators of organic material which they make from mineral salts and carbon dioxide in the presence of sunlight and chlorophyll. Animals cannot make organic material, so they feed on plants or on other animals which do so. Such a relationship is known by ecologists as a food chain.

The purpose of this book is to illustrate and describe the forms and habits of some of the world's marine plants and animals, and to relate them to the environmental conditions imposed by the seas. As in other areas of science, man's knowledge of the sea and the organisms it contains is constantly developing. Marine biology as a scientific discipline is, of course, relatively new, although many animals living in the sea were described by Aristotle. Centuries later, great expeditions were mounted by the expanding nations of Europe, such as that of James Cook to Australia and New Zealand in the eighteenth century. The results of investigations by naturalists in association with these voyages of discovery, set in the context of a scientific awakening at home, stimulated people like the Swede, Carl Linnaeus, to embark upon 'Systema Naturae'. This is a founding work of botany and zoology which attempts to classify all living things. As a result, by the middle of the nineteenth century many 'naturalists' were carefully describing and naming a flood of organisms found to be new to science. These developments in biology, inspired men such as Charles Darwin and Alfred Russell Wallace to develop the theory of evolution. Their pronouncements on the idea of evolution stimulated the need for experimental biology. One area to benefit from this stimulus was marine biology. Laboratories and research vessels began to appear which were to transform man's understanding of the natural world in the sea. As the descriptions of species became more complete, so investigations into the habits of plants and animals and their mechanisms of dealing with environmental pressures were able to commence, and at the present time much analytical research is under way.

Throughout history the sea has been of profound significance to man, and especially so to those cultures and civilisations living near the coasts. It has provided him with a source of food, a means of transport and a way for commerce. Because of scientific and technological advances, the twentieth century has seen the emergence of new areas for marine exploitation. These include intensive fishing, gravel dredging and gas and mineral extracting. The seas are rich in many forms of resources, but most strikingly they provide habitats for many beautiful but poorly understood plants and animals. Although this book indicates that much of the biology of some organisms is well understood, many scientific questions relating to them remain unanswered.

Seashores

Rocky shores occasionally have quite large inhabitants. The Marine Iguanas, like sentinels from another age in the Earth's history, are a feature of the volcanic shores of the Galápagos Islands.

The rocky shore

When you next visit a rocky shore, take a walk from the top of the beach down to the water's edge. If the tide is out you will notice that the plants and animals are arranged in belts or zones. From a distance about four zones can be made out which are often sharply defined by colour. At close quarters these colour bands will be seen to relate to the dominant organisms. Such zones are characteristic of tidal shores the world over.

Why should life on the shore be zoned? The seashore is unique among habitats as it is covered and uncovered by the tides once every twelve and a half hours or so. It extends from the highest point to which the tides flow down to the lowest point to which they ebb, so it consists of a relatively narrow fringe running round all the land masses. Its extent is not governed merely by the vertical rise and fall of the tide but by the amount of exposure to prevailing weather and by the slope of the land which forms it. As the tides rise and fall a gradation of conditions is created starting from those at the top of the beach which are exposed to air for most of the time, to those at the bottom which are largely submerged. Because different species of animals and plants are able to tolerate different degrees of immersion and desiccation, they can take up life in different positions. Other factors such as competition for space and food also play a part.

On the shore, therefore, physical factors – wetness and dryness, wave action and temperature – are in a continuing state of flux.

Tides themselves are caused by the gravitational effect of the sun and the moon pulling upon the waters of the ocean. The degree of water movement, however, does not end simply with the rise and fall of the tide. At certain times of the lunar month the sun and the moon pull together producing a greater rise and fall (the spring tides). At other times their pull is restricted because they are exerting a force at right angles to each other, so that

Right
A shingle shore such as this, the Chesil Beach in Southern England, shows damp marks left by the receding tide. Very few animals can tolerate the harsh conditions of shingle so there are no signs of zoned life.

Below
A rocky shore immediately gives the impression of flourishing life. Here zonation of yellow and black lichens, grey barnacles and brown seaweed can be clearly seen.

Many sea-snails such as this top shell can close the aperture of their shells with a horny operculum attached to the rear of the foot. Closure prevents desiccation at low tide.

the rise and fall is reduced (the neap tides). If we look carefully at the lunar month, we find that the cycle of twenty-eight days includes two spells of spring tides and two spells of neap tides arranged alternately, the intervening tides forming gradations between them. Exposure to different winds can bring about variations on a shore, as they can also affect the behaviour of the sea.

Since each tide follows its predecessor by approximately twelve and a half hours, the time of a low or high tide advances throughout the month. Consequently, the times of immersion or exposure to air of the seashore become later each day by about half an hour returning to the original time after about a fortnight.

Exposure to air and sun in midsummer may cause serious dehydration and physiological stress in an organism, which at other times is bathed in sea water, and the temperature of which normally changes only a few degrees each

month. Evaporation can increase the salinity of the sea water held within the animal's body. Rain, on the other hand, may dilute it. On winter nights at low tide water on the shore could freeze, although the sea itself only very rarely does so. There are other points. Waterborne oxygen and food particles will be available at high tide. Predators may move in from deeper water at high tide when locomotion will be easier, because of the support provided by the water for large animals with denser bodies or shells.

As we approach the edge of the rocks, the land plants thin out and the rocks themselves become covered by small encrusting growths called lichens. Lichens are a biological curiosity, being half fungus and half alga. Because of their black or yellow pigments they often create strong colour bands, and although they are not covered by the sea at high tide, they normally grow under the influence of spray.

The lichen zone. This is the uppermost zone or belt of the shore and a variety of small insects and other arthropods like centipedes may live here. Any crevices in this zone may be inhabited by animals whose affinities are divided between land and sea. One such is a group of arthropods commonly known as

sea-slaters. These resemble large woodlice and are effectively terrestrial but require a certain amount of water to live. Also in the cracks and among stones are many small sea-snails known as periwinkles. One family of these, the Littorinidae, has representatives on most rocky shores of the world and is of particular interest to ecologists because its different species have modified themselves to live at different levels on the shore. Up to four species may be found in some cases between the lichen zone and the bottom of the beach. Periwinkles are all herbivores and they scrape their plant food, like lichens and algae, from the rocks using their toothed tongue (radula). Some species can breathe air and some breathe water. Some species have a planktonic larva, while in others the larval stage is passed through in the egg case so that the juveniles are released into the same zone as their parents. The smallest species are characteristic of the lichen zone. Film-like growths of red and blue-green algae grow at the bottom of the lichen zone around the high water mark of more sheltered shores. They give the rocks a slippery texture. Some green algae may be met in pools of stagnant water at this level.

The barnacle zone. Barnacles are the dominant animals of the second zone. They also belong to the great phylum, Arthropoda, like the sea-slaters but are far from typical members. In fact it was not until the discovery of the barnacle larva, which does look like a proper arthropod, that these animals were correctly classified. Before they had originally been mistaken for molluscs.

After feeding on plankton the swimming barnacle larva settles on the rocks and metamorphoses to an adult. Its head becomes cemented down so that once a site is chosen it cannot be changed. The body is encased by four or six strong greyish plates which protect it from the force of the waves. The appendages which would have formed the legs develop as fine branching structures which can be protruded from the shell like a fine basket. When covered by water this basket is used to filter food particles from the sea, so the animals can only feed when covered by water. Because they cannot move they may be exposed to extremes of temperature. By retaining some water in the shell, they can allow this to evaporate slowly at low tide thus cooling themselves on hot days.

Limpets are also familiar animals from the barnacle zone. They are essentially sea-snails with uncoiled shells. Like the garden snails they creep around on a well-developed muscular foot. Unlike many other snails of the shore they cannot withdraw their bodies because the uncoiled shell is not long enough. Therefore, for defence, they depend on adhering to the rocks which they manage to do by means of their powerful sucker foot. In this way only the strongest predators can prize them away to get at the soft body within the shell. When the tide is in, limpets move from their resting place, often marked on the rocks with the imprint of their shell, and browse on fine algal growths. They appear to return to the same spot when the tide ebbs. They use a toothed tongue or radula, like the periwinkles, to scrape the food from the rock surface.

Towards the lower extremity of the barnacle zone, some conspicuous seaweeds may be encountered. In the north Atlantic one such is *Fucus spiralis* or the Spiral Wrack. In the lower part of the shore its close relations the

Above left
A rock pool, like this one in the barnacle zone, interferes with the normal zonation pattern because it traps the tide. Notice how the barnacles end abruptly where the water starts and how red encrusting algae develop in the pool.

Left
Barnacles cluster tightly on the rocks of shores in many parts of the world. Here *Eulalia viridis*, a green polychaete worm, creeps between the barnacle shells at low water.

Above
The grazing activities of herbivores like limpets limit the activity of encrusting algae. The limit of feeding activities is marked by the appearance of algal lawn as shown in the foreground.

Serrated Wrack *(Fucus serratus)* and the Bladder Wrack *(Fucus vesiculosus)* will frequently be met. Wracks, like most of the larger algae, are attached to rocks by means of well-developed hold-fasts. Their fronds are covered by a slimy secretion which lubricates them as they swash back and forth with the waves, and this prevents some abrasion of their tissues. This secretion also helps to reduce water loss by evaporation in strong sunlight when the tide is out. A variety of small animals rely on the shade of the wracks for protection from sunlight and water loss at low water. Other animals feed on them, so they provide opportunities for many species.

The mussel zone. The bottom of the barnacle zone is marked by the start of the mussel zone, which normally appears as a clear bluish band across the beach, and whose upper limits are normally apparent as a well-defined line. The plants and animals in this zone will be covered by sea water for about half of the tidal range. Mussels, like the limpets and periwinkles, belong to the great invertebrate phylum known as Mollusca, but to a very different sub-group: that which includes all the Mollusca with two shells. Along with the scallops and the oysters and their relations, they are known as bivalves and as such, they lead a very sedentary life. For most of the time adult mussels lie attached to the substratum by means of filaments they secrete. The filaments are described by scientists as the byssus; the cook preparing the mussels for the table calls them the beard. The byssus can be broken from the mussel and a new one produced if necessary. The entire body of the mussel is enclosed in the paired shells. These hinge at the top of the body, so that they open on the animal's underside. The gills which lie just inside are equipped with vibrating

microscopic filaments (cilia) which pump in water to bring oxygen and remove carbon dioxide. At the same time, this water brings a stream of food particles, for, like the barnacles, the mussels are filter-feeders. Indeed, they could not feed in any other way because they cannot move about enough. The water enters and leaves by different routes and the suspended particles are filtered out as they pass through the gills. Particles of the wrong size or texture are rejected, but the acceptable ones are passed to the mouth and ingested. Despite their shells, mussels have a number of predators which are good at breaking through their defences. Some birds like the Oystercatcher (*Haematopus ostralegus*) stab the shells at one particular point with their sharp beaks and break them open so as to get at the soft flesh inside. Starfish use a different method. They climb on top of the mussel and lever the shells just apart with their multitudes of sucker feet. A slight gap of a millimetre or so is sufficient for the starfish to insert its stomach membranes, which it folds out through its mouth,

into the shell of the bivalve. The digestive juices then start work, rapidly killing and digesting the prey on the spot.

Mussel shells themselves form a suitable habitat for various other small organisms, including young crabs and free-moving worms. Of course there are other animals grazing on the rocks in this region as well. On fairly steep shores we may find another species of limpet and more barnacles, again of another species. The animals in this situation need to be covered by water more of the time than their upper-shore neighbours. Also in this region one usually encounters several types of red algae. In all algae there is a basic green pigment, chlorophyll, which is essential for photosynthesis. In the brown species this is masked by an additional brown pigment, and in the red algae by an additional red one.

If the shore is not steep at the mussel zone, algae may be very abundant, and thick growths of fronds can cover both attached and free-living animals when the tide is out. In the mussel zone we frequently encounter small- to

moderate-sized predatory whelks. These animals live surrounded by their prey which frequently takes the form of barnacles. If one searches among the barnacles it is usually easy to find some empty shells whose occupants have almost certainly fallen victim to the rasping radula and sucking mouth of the Dogwhelk (*Nucella lapillus*). When these whelks themselves die, their shells are frequently taken over by small hermit crabs, whose soft abdomens are poorly protected by their own hard external skeleton.

When the tide is out there will be a number of conspicuous blobs of coloured jelly visible on the rocks. These are sea-anemones. When the tide returns to cover them they will open their bodies to reveal flower-like arrangements of tentacles with a mouth set in the middle. Sea-anemones are familiar to many people and they are quite common shore and rockpool in-habitants. They are attached to rocks by a basal sucker-disc which they are able to move and progress slowly from one place to another. They belong to the phylum Coelenterata, whose members are

Left
Seaweeds on the upper and middle shores
have to withstand exposure to air and sun
as the tide ebbs. *Enteromorpha* shows
green chlorophyll pigment unmasked by
brown, as is the case in *Fucus spiralis.*

Right
Periwinkles are important algal grazers.
This species, *Littorina obtusata,* is often
found on the Bladder Wrack *(Fucus
vesiculosus)*. It is said the flattened shells
mimic the bladders of the wrack, thus
reducing danger from predators.

Below
Some small gastropods like the Dog Whelk
(Nucella lapillus) live in the same zone as
barnacles and prey upon them. The small
vase-shaped egg cases of the Whelk can
also be seen.

considered to show one of the simplest levels of organization in the animal kingdom. The sac-like body is made up of two cell-layers, whereas in most other types of animal three layers are involved. The mouth is set in a disc forming the top of the animal and the sucker provides the disc for the base of the body. The whorls of tentacles which surround the mouth are used for capturing the prey and moving it into the mouth. To do this more efficiently the tentacles are armed with stinging cells which act upon contact with small prey. They also serve as a means of defence. Digestive waste is passed out through the mouth which also serves as an anus. Some anemones are able to reproduce by splitting into two; others reproduce sexually and form a swimming larva; some can use both methods. Among their few predators are the highly coloured sea-slugs. These distant relatives of the garden slugs are not deterred by the stinging cells, and when pieces of anemone are bitten off, the slug's digestive system sorts out the stinging cells from the rest of the food, and allows them to be

dispersed around the skin of the sea-slug where they can be used secondhand to defend it!

The shore crab is a familiar crustacean of the mussel zone. These crabs are frequently camouflaged to make them relatively inconspicuous among the seaweed. They are, however, well able to defend themselves with their pincers and they can shuffle quite rapidly across the rocks. Sometimes these crabs can be infected with a type of parasitic barnacle which grows like a lump on the folded abdomen. This parasite secretes its own hormones which, in effect, castrate the crab; but since the crab can usually outlive its parasite, sexuality normally returns. The crab may be encountered carrying its eggs under the abdomen. The eggs have a more granular appearance than the parasite, which is covered with smooth skin.

The kelp zone. The lowest zone of the rocky shore ends at the extreme low water line and stretches to that point from the end of the mussel zone. It is characterized by the presence of red algae and large brown algae (kelps). These plants

need to be covered by the sea for most of the time, but they may be encountered further up the shore in rock pools. One group of algae has a characteristically chalky deposit associated with it. An example of this is the genus *Corallina* with its fine chalky branching fronds. These fronds are still flexible because each little section is jointed to its neighbours. Quite different in appearance and texture are those coralline algae which have massive chalky bodies. The genera *Lithothamnion* and *Porolithon* are examples of these, with species distributed all over the world. They are mainly found as encrusting forms. Similarly the kelps of the genus *Laminaria* and its relatives have representatives in all temperate waters but brown algae, unlike some of the red species, do not flourish in the warmer seas. While the red algae are frequently small plants, the brown algae can be quite large and some of the kelps are enormous. Kelps usually abound at the bottom of the lowest zone on the seashore, but they are often uncovered only at the lowest tides. Such algae form the famous kelp forests off the coast of California.

16

Above left
When the tide ebbs, the beadlet anemones withdraw their tentacles and close up their bodies. A thin layer of slime helps to cut down water loss.

Above
Crabs are common members of the rocky shore fauna; they feed on a variety of organisms. Some tropical species like this one may be conspicuously marked.

Right
Red algae are characteristic of the lower shore. The green colonial anemones in the foreground show how the tentacles surround the mouth of each individual when it is open.

A wide variety of animals is to be encountered on the lower shore, and they are generally those which have strong affinities with the truly marine species. A number depend on the weeds for an anchorage or for shelter. Kelps, in particular, support a diverse fauna. On the fronds we may find colonies of hydroids, such as the genus *Obelia*. These little coelenterates are distant relatives of the sea-anemones and use tentacles with stinging cells to capture their microscopic prey. Unlike the anemones they secrete a fine horny skeleton which supports and protects their bodies. They also live in colonies of many individuals rather than singly. Somewhat resembling the hydroids, but zoologically very different from them are the moss animals or sea-mats. These are colonial too, and they live as many individuals grouped together.

As with the hydroids, the overall shape of the colony and the form of growth depends on the species concerned and where they grow. The sea-mats belong to the phylum Bryozoa, and they have either horny or chalky skeletons surrounding each individual. They grow well on the fronds of kelps and wracks and on rocks and shells both on the lower shore and in the shallow sea. They live by filtering water through

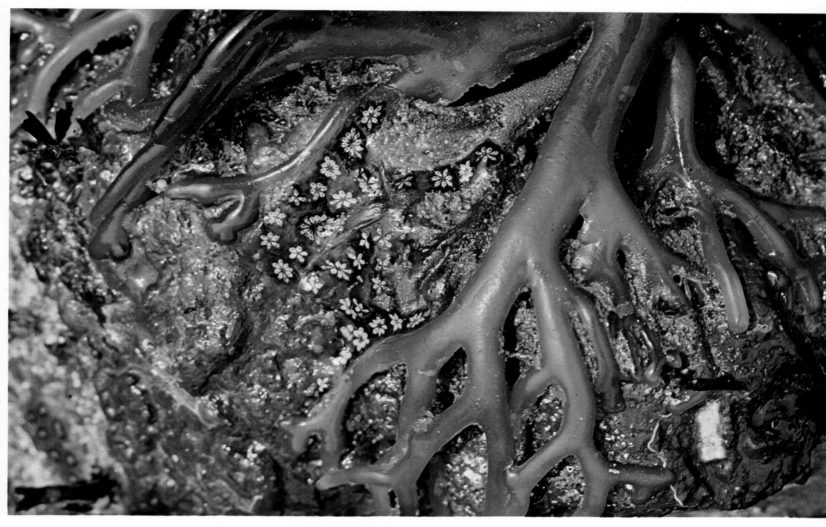

Left
Extreme low-water spring tides reveal the holdfasts and stalks of kelps in the lowermost shore zone.

Above
Kelps require firm attachment organs to anchor them in all weather conditions. These holdfasts make ideal habitats for small organisms such as this orange sponge and the green and white 'star' or compound sea-squirt (*Botryllus*).

Right
Most starfish, such as *Pisaster ochraceus* from Puget Sound, Washington State, USA, locate their food by means of waterborne chemicals emitted from their prey. They then creep towards their food by means of coordinated stepping movements of their numerous tube-feet.

A number of small fish live under stones and in rock pool crevices. The Common Blenny (Blennius pholis) is a ubiquitous member of the European rocky shore fauna. Its cryptic coloration provides good camouflage.

their tentacle crown which can be pulled back into the body when danger threatens. A mouth opens in the middle of the tentacles and leads to a U-shaped intestine which terminates in an anus situated just outside the ring of tentacles. There are no stinging cells. Some of the bryozoans with chalky skeletons have proved economically important because their skeletons, like those of some tube-worms and barnacles, foul ships and industrial marine plant.

Some fish may be found at the bottom of the lower shore. The blenny family has many representatives as does the goby family. Other groups such as wrasses and butterfishes may occur here and in rockpools, and the general abundance of animal life in this zone makes it a rich hunting ground for some seabirds.

The sandy shore
After walking across a rocky shore at low tide it makes an instructive exercise to do the same on a sandy beach. In this way direct com-

Above
From a distance, sandy and muddy shores may appear devoid of life. As this picture shows, the presence of solid objects, such as piles, immediately provides a substratum for barnacles which would otherwise be unable to colonize the area.

Left
Surface features such as the casts of the lug-worm *(Arenicola)* are an indication of the density of hidden life in sandy shores.

21

Prawns are frequently found trapped in rock pools where they scavenge for food.

parisons can be drawn between the two. One's first impressions of a sandy shore under these circumstances are of barrenness. Whereas on rocks one can immediately distinguish belts of weed and zones like the mussel zone, this is not possible on sand because sand does not provide a stable attachment medium. If one looks very carefully, the sand's surface will nevertheless be found to show signs of life below in the form of worm casts, tubes and burrow entrances.

The animals of the sandy beach live mainly within the substrate. To do this they must be able to burrow and carry on life's processes below the surface of the sand. Relatively few types of animal have evolved to cope with this situation, so one will not encounter the diversity of form that is apparent on the rocky shore. On the other hand, in the right kind of sand quantities of sand-dwelling animals may be very high. Algae do not grow well in sand as they have no roots to anchor them down. A few plants, the sea-

grasses, for instance, which are quite closely related to the land-grasses do flourish under favourable circumstances. Sea-grasses may be encountered on the lower shore and in shallow water where they often form distinct and peculiar communities. Their presence tends to stabilize the sand and allow other organisms, particularly animals to come in and colonize it. While the tidal factors already discussed play an important role in governing sandy shore life, the interactions of some of the other physical factors may be somewhat modified. Radiant heat from the sun does not penetrate a long way into the sand, fresh water from rain or streams runs over the top and the grains themselves tend to hold water by capillary action when the tide is out. All in all, the sandy shore is an excellent habitat for those organisms which can burrow and extract food, either by digesting the organic matter contained in the sand itself, or by filtering sea water for suspended particles. The communities of a sandy shore are, in fact, zoned but their burrowing habits hide the marks of zonation.

The upper zone of the shore, especially round the strandline, is often characterized by a number of arthropods such as flies and sand hoppers. Sand hoppers of the genus *Talitrus* feed on organic matter decomposing amid the flotsam of dead seaweed and animals so often washed up. These animals live within the debris and resemble woodlice which have been flattened from side to side rather than from top to bottom as is the case with the sea-slaters.

Further down the shore, towards the mid-zone, two distinct types of animal may be found. First, there are those which actually live in the sand and which, if they move at all as adults, do so only slowly. Second, there are those animals which are highly mobile surface-dwellers and which retreat to the sea when the tide ebbs and return when it flows. These are, strictly speaking, not shore animals at all and belong to the shallow sea, but they do some-times get left behind and become stranded at low water. They include flatfish, cuttle-fish, swimming crabs, prawns and shrimps and purely

The siphons of the cockle enable water to enter and leave its body. The shells are secreted by the mantle, the edges of which can be made out in the open gap.

floating animals, like the sea-gooseberries which will be described later.

We have already seen some of the problems faced by animals which live in the sand. How are these overcome? The answer lies in the extensive adaptations which have arisen during the course of their evolution. They must be able to burrow and to maintain themselves under the sand. In the case of many of the worms this is relatively easy for they can dig themselves in either by changing their shape or by using the small paddle-like extensions of their bodies known as parapodia. Some worms, like the Sand Mason *(Lanice conchilega)* live in the middle shore and construct tubes in the sand using a special cement secreted by the body, into the walls of which sand grains are set. This confers strength and support so that the worm is protected under the sand and supported when it rises in the tube to collect food particles at high water, which it does with the aid of long filamentous projections at the head end.

One group which is well represented in the sandy environment generally is the bivalves. Bi-valve anatomy is complex, but the underside of the body is developed into a foot which corresponds anatomically to the creeping foot of their distant relatives, the gastropod snails. In some of the non-burrowing bivalves, like the mussels, it is the foot that secretes the byssus. In burrowers this organ is important for burrowing. It becomes extended down and penetrates the sand, expanding to form an anchor. Then the upper part contracts pulling the shells below the surface. This operation may be repeated many times until the animal has reached the required depth. The slim shells of the tellins from the middle and lower shore and the narrow razor-shells lend themselves well to this mode of life.

Getting under the sand is but one problem. Food and respiration are two more important ones. Because marine animals are dependent on a supply of water, bringing oxygen and suspended food, such a supply must be maintained even under the sand. Most bivalves overcome this problem by using siphons, tubular extensions of the body which reach up to the surface. There is an inhalant siphon and an exhalant siphon. In the tellins themselves the two are separate and the inhalant siphon is longer. The animal uses it to suck up food particles deposited on the sands surface. In the razor shells the two siphons are united but they function similarly although only suspended particles rather than deposited ones are brought in with the water. Waste products and carbon dioxide in solution leave via the exhalant siphon as do the sperms and eggs in the breeding season.

Burrowing crabs such as the masked crabs show amazing adaptations. Their two antennae are each shaped like pieces of guttering so that when held side by side a tube to the surface is formed down which a supply of water can be pumped. However, these crabs have to emerge to feed using their pincers to catch prey. Masked crabs are usually found on the lower shore along with the well-known lug-worms from the genus *Arenicola*. The lug-worms have succeeded without many structural modifications but they show considerable physiological ones. The animal feeds like the earthworms by ingesting the substrate as it burrows through

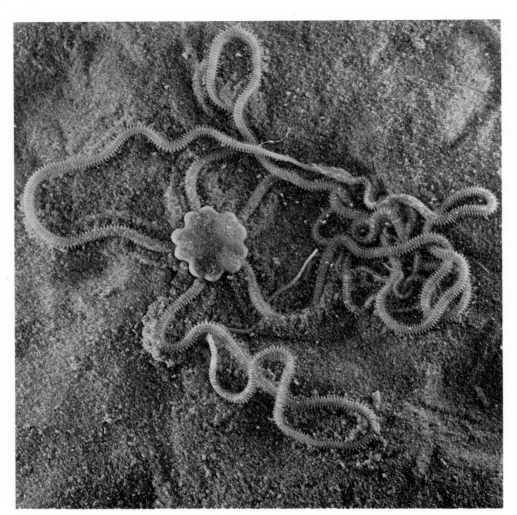

and removing from the sand any organic matter that it may contain. 'Cleaned' sand is passed out from the anus. As it burrows, a current of water is drawn through the burrow in the opposite direction to provide a supply of oxygen.

There are other approaches to life in the sand such as are shown by the predators. A good example is the predatory Necklace Shell (*Natica alderi*), a very distant relative of the periwinkles. This gastropod mollusc burrows with the aid of its expanded foot. When it encounters its prey, for instance a bivalve, it drills into the shell with its toothed

radula, and sucks out the body contents using an extended tubular mouth. Another strategy for predators on the lower shore is adopted by the burrowing starfish which engulf whole any prey they come across in the sand. They have a good sense of smell which helps them to locate their prey. Yet another is the 'sit and wait' strategy of the burrowing sea-anemones whose bodies are so formed that, unlike their rocky shore relatives, they do not depend on a hard substrate for attachment. They wait, tentacles extended, for prey to stumble into their deadly grasp.

Sand flats and muddy shores

Sandy shores are not the only types of soft substrate encountered between the tides. Many hundreds of square kilometres of sand flats and muddy shores and estuaries exist throughout the world. Mud or silt particles are much finer than sand and, consequently, hold the water better at low tide. Also, near river mouths they retain much more organic matter so they can provide food for a dense fauna and may be very productive. The sorts of animals encountered here are very similar to those of the sandy shore. Bivalves, like cockles and clams, flourish;

Left
The head of this omnivorous ragworm *(Nereis)* can be seen to be equipped with sensory antennae and palps. Note the well-developed paddles or parapodia along either side of the body.

Below
Eel-grass *(Zostera marina)* forms an important substratum for hydroids, anemones and bryozoans which are unable to settle directly on the sand or mud.

Right
The Oystercatcher *(Haemotopus ostralegus)* waits in flocks for the tide to ebb and expose the animals of the lower shore on which it feeds.

Corophium volutator is a small amphipod (sideways flattened) crustacean which burrows in mud. It feeds on detritus and its abdominal appendages pump water through the burrow, whose walls are strengthened with a sticky secretion.

worms like the ragworms *(Nereis)* and the bamboo worms *(Maldane)* burrow in the mud, and a variety of small burrowing arthropods may be encountered.

Because of all the animal life present, these shores may support a number of predators. Birds like oystercatchers and gulls usually occur in great numbers. Where there are extensive communities of sea-grasses and reeds, geese and other herbivorous birds may roam.

If there are plenty of fish these can support a colony of seals. A number of estuaries and mud flats support fishing and shellfish industries and are therefore commercially important. Sadly, too, they are often badly polluted by industrial developments. Although when compared to rocky or sandy beaches they may be considered unattractive, these areas are frequently rich in nutrients and support a wealth of invertebrates and bird life.

Shallow seas

The shallow seas may be loosely described as those which overlie the continental shelves. They form the so-called neritic province, in contrast to the oceanic provinces of the world. In general, their depths range down to about 150 metres (492 feet) as the seabed slopes gently away from the land masses. All the continents of the world are surrounded by a continental shelf whose width may vary from 1 or 2 kilometres (0·6 and 1 mile) to 1,000 kilometres (600 miles). To seaward, as its outermost limit, the shelf gives way to the continental slope which descends more steeply to the floor of the ocean. The North Sea is an example of a sea which lies entirely between land masses. It is relatively shallow for the whole of its area since it lies above the continental shelf stretching between Britain and Europe.

In biological terms the shallow seas are generally very productive. This means that they have the potential for producing a lot of organic material, such as fish or plankton, from inorganic matter. There are various reasons why they are more able to do this than the oceans. Firstly, they are under the influence of the land, at least in the inshore waters. The land provides a wealth of inorganic matter in the form of salts dissolved in fresh water leaving the rivers. Some organic substances can also be washed out to sea. The inorganic substances are most important for the nutrition of phytoplankton, and what is naturally present in sea water is augmented by the supplies from land. Phytoplankton may then flourish

according to the season and provide the basic stage in the complex food chains of the sea. The phytoplankton are fed on by herbivorous plankton such as copepods, which are then in turn preyed on by carnivores like fish larvae and invertebrates like sea-gooseberries and arrow-worms.

The second point relates to the depth of shallow seas. Because there is relatively little volume of water, the water volume/surface area ratio is low compared with the oceans. Here we must pause to consider some physical characters of sea water itself in relation to living organisms.

A great variety of worms live in the shallow seas. *Aphrodite*, the Sea-mouse, is not very worm-like in appearance. It creeps on its flattened underside. The upper surface is protected by scales and iridescent hairs.

In the last chapter it was pointed out that plants are the primary producers of organic material. Only they can elaborate organic molecules from inorganic ones by means of photosynthesis. For this process they use sunlight and the green pigment chlorophyll. On the shore and in very shallow water large algae grow and carry out this process. Away from the coast, either in the shallow sea or in the open ocean, phytoplankton convert mineral salts and carbon dioxide into organic matter in the form of sugars and plant protein. These microscopic organisms are the basis for all other life, either directly or indirectly. For life in the shallow sea, therefore, these organisms are very important.

Because of the need for sunlight before photosynthesis can take place, phytoplankton tend to swim upwards in the water when illuminated and sink down in the dark. In many shallow seas there are plenty of mineral salts in the surface layers which help the phytoplankton to be productive; hence the whole sea is productive.

Physical conditions change throughout the year so that the availability of mineral salts is re- duced under certain circumstances. After the winter storms the sea is well mixed up but as warmer, calmer weather sets in, the surface waters may get hotter than the layers below. When this happens a less dense warm layer floats on top of a colder heavier layer, and the point where they join (the interface) becomes well defined. This inter- face is known as a thermocline. If one were to don a diver's suit and descend through the thermocline, one would experience a strange sensation of sudden cooling as one passed down into the lower layer. One important fact about the thermocline is that when it becomes established, there is hardly any communication between the two water layers. Consequently, as the mineral salts in the upper layer are used up by the phytoplankton, they are not replaced by diffusion from the deeper water as they would be at times of the year when there is better mixing of the water. This results in starvation of the phyto- plankton and a reduction in their population which in turn affects all other planktonic animals which de- pend upon them.

The hotter the climate the more marked the thermocline may be.

In a warm sea, such as the Red Sea, the thermocline may be very marked and the absence of much rough weather may mean that it is an almost permanent feature throughout the year. This means that the productivity of the sea may be relatively low in comparison with colder stormier waters.

While we are not dealing specifically with planktonic life in this chapter, planktonic factors have a profound effect on the food or energy supply to all members of the shallow-sea community. These effects register on the lives of larvae and the lives of adults. In the last chapter we saw that the shore communities were governed directly by environmental pressures, and that each type of environment had characteristic fauna and flora. The same is true in the shallow sea. The form of the shallow seabed (substratum) for example, rock, shingle, shell, gravel or sand, directly determines what form of life we shall find there. Further, the degree of exposure, the tides and the prevailing

currents all make a difference. Unlike the shore species, shallow sea organisms will not need to withstand exposure to air or to wide temperature ranges. For the most part they will experience constant or gradually changing conditions of salinity, temperature, food and oxygen, according to the seasonal influence.

There will, however, be some different problems such as sedimentation near river mouths and sand banks with which they may have to contend. Sediments can smother attached organisms which cannot move to clear themselves. For some animals, though, sediments may form a rich food supply.

Let us look at some members of a typical rocky bottom community. Sponges are characteristic of many shallow seabeds and they will grow wherever there is a suitable substratum and waterborne food supply. They are very simple animals with about three cell types in the body. Essentially they are shaped like a vase, or a number of vases

Above
Zoanthids are colonial sea-anemones which in most respects resemble their solitary relations. Notice how the individual polyps arise from the common mass and have the typical features of a central mouth surrounded by a ring of tentacles. These specimens encrust stones and rocks and come from New Zealand waters.

Above right
The precious red coral of the Mediterranean is not a reef-builder, nor is it very closely related to reef-building corals. Growing with it are some brilliant yellow anemones.

Right
Chaetopterus variopedatus is probably the most specialized annelid worm. It secretes a parchment-like tube and pumps water through this by means of three large thoracic paddles. Food particles are trapped in a mucous bag.

Overleaf
Eupagurus bernhardus, like many species of hermit crab, is often found associated with sea-anemones. It is thought that they protect the crab from the attacks of octopuses and, in return, share the crab's food.

34

side by side. The sides of the vase-like body are perforated by hundreds of small holes through which water is pumped by cilia. Any suitable food particles are trapped and ingested by the cells at this stage. The water brings with it oxygen and removes carbon dioxide. It passes from the small holes to the central cavity of the sponge and leaves via the large aperture at the top. Some sponges secrete particles of chalk or silica in the walls of their bodies which help to support them. In others, horny fibres do the job. It is types of horny sponges which are used for bath sponges after the live tissue has been cleared away. Because there is a good supply of water, food and oxygen inside the sponge, and somewhere to hide from predators, other animals may make their home there. Marine worms, crustaceans and small fish may be found within the central cavity.

Various types of sea-anemone abound in the shallow seas and, freed from the problems of shore life, more species occur here. Interesting examples are those which live in association with other organisms such as hermit crabs (family Paguridae). Various species of hermit from rocky and sandy bottoms associate with anemones which live attached to the gastropod shell used by the crab. It is thought that the anemones collect quite a lot of food from the crab's messy feeding habits, and that at the same time the crabs acquire protection from predators on account of the anemones' stinging cells. Close relatives of the anemones are the true or stony corals. In many temperate parts of the world solitary corals can occur attached to rocks or stones. Like anemones they depend on capturing prey with their tentacles.

Many animals, including the corals, secrete skeletons or tubes and shells from calcium carbonate which occurs in solution in sea water. The shells of barnacles and those of molluscs have already been mentioned. Some of the sedentary worms secrete calcareous tubes and these frequently grow encrusting submarine rocks and boulders. These worms are members of the annelid phylum belonging to the family Serpulidae. They have a well-defined tube and a body perfectly modified for life within it. A special region behind the worm's head, known as the collar, secretes the calcium carbonate which forms

Other tube-dwelling annelids filter feed by means of a crown of tentacles which serve for respiration too. When danger threatens they can withdraw completely into their tubes.

the tube. The head itself is somewhat reduced and bears a number of filamentous tentacles. When feeding, these tentacles are extended from the mouth of the tube to trap floating food particles and micro-organisms. Should danger threaten, the worms are sensitive to shadows and changes of light intensity caused by movements of other animals, the tentacles can be withdrawn and a little stopper pulled in afterwards to close the entrance to the tube. The tentacles also provide the surface for gas exchange, essential for respiration.

Many other organisms encrust the rocks apart from worms and corals. A great variety of bryozoans related to those we have met on the kelps will occur. Some of these have chalky skeletons, too. There will be different types of barnacles, often very large ones which can grow rapidly due to the availability of a constant food supply. Various molluscs live on the rocks including some bivalves which actually bore into the stone itself by mechanical or chemical action. In this way they are protected from predators like starfish. Gastropods such as whelks, top-shells and sea-slugs flourish

where there are plenty of sedentary or slow-moving animals for them to prey on.

Lurking among the rocks we may also be able to discover that most highly evolved mollusc, the octopus (Octopus). This animal is well known to zoologists for the size and capabilities of its brain. The octopus is a very beautiful animal and one that is capable of changing its coloration in seconds according to the background over which it passes. In this way it escapes notice by its potential prey as it stalks up. Also it is protected from the attacks of possible predators who will experience difficulty in spotting it. The ability to change colour is quite widespread among marine animals and is shown by fish and crustaceans too.

By means of stealthy or rapid movements the octopus can overtake prey such as crabs and lobsters which it seizes by means of the suckered arms. The prey is then bitten with a parrot-like beak and injected with poisonous saliva which quickly renders it senseless. In this way the octopus can quickly overcome quite powerful and active prey.

Above
The reproductive bodies of this tubularian hydroid are dark pink in colour while the feeding tentacles are pale. Notice the small amphipod crustaceans (flattened sideways) scavenging detritus on the stems.

Left
The octopuses secrete poisonous saliva which they inject into their prey. The saliva of the beautiful Blue-ringed Octopus from Australia is so poisonous that it can kill a man.

Overleaf
The mantis-shrimp *(Squilla)* has a powerful abdomen and tailfan. By flicking its body it can swim backwards rapidly to escape predators. It lacks nippers as such but captures its prey in the 'elbow' of its front legs.

The Atlantic Lobster (*Homarus gammarus*) has two powerful claws. Notice that the left-hand claw, as you look at the picture, is stouter and more rugged. This is used for crushing the prey, while the right-hand one is adapted for snatching food.

The lobster may be thought of as being well armed with its thick shell and powerful pincers. If these two pincers are examined closely they will be found to be different in form. Generally, though not always, the right pincer is the larger and stouter. This is used for smashing shells and crushing victims, while the left one is used for grabbing at moving objects. If the right pincer is lost in an accident the left one develops into a bigger smashing claw, while the right regenerates as a completely new grabber. Regeneration of lost parts is a characteristic of many invertebrates including starfish, arthropods and annelid worms. The ability to replace and repair lost parts appears to decline the higher one goes in the animal kingdom.

Still among the animals of the rocky bottoms, mention must be made of one other interesting group of invertebrate animals. These are the sea-squirts (*Tunicata*). Sea-squirts live attached to rocks and the holdfasts of weeds such as kelps. They consist of a sac-like outer coating, made of cellulose, called a tunic, within which lies the large pharynx. Water is pumped into the pharynx by the action of millions of cilia which line its walls. Particles of food are once again filtered from the water and passed into the twisted gut. Filtered sea water leaves by a second opening. The head of the adults is virtually non-existent, but the larvae, which swim free in the plankton, slightly resemble the tadpoles of amphibia. It was only by studying the larvae, as with the barnacles, that the zoological affinities of the sea-squirt became apparent, and these creatures are now known to be related to the animals with backbones.

What of the plants of the shallow seabed? Generally they are absent

below the depth to which light penetrates. In shallow water and where the water is clear and free from sediment, some may be abundant. This is particularly true of the kelps. It is also true of a number of species of small red coralline algae like *Lithothamnion* and *Porolithon*, which under certain circumstances can grow prolifically in both temperate and tropical waters competing with some of the attached animals for space on the rocks. Whenever certain types of plant grow in abundance, they usually dictate what type of animal can live among them. Thus, it is that distinct communities come to be associated with certain types of weed, for example, *Sargassum*, kelps or the coralline algae. Remarkable as it may sound, some of the bryozoans, such as *Flustra*, grow in a plant-like fashion and they have a distinct community of other animals associated with them too,

including sea-spiders which specialize as predators of the bryozoan animals.

Little has been said so far about fish, either amid rocks on the shore or in the shallow sea. Needless to say, many species may be found there. Where plant life abounds some may be encountered that feed specifically on weeds, such as mullet (Mugilidae). In tropical waters, which will be discussed later, many herbivorous types such as surgeon-fish (Acanthuridae) inhabit reefs. A rocky bottom provides good cover for territorial fish, that is fish which stay largely in one place like certain wrasse (Labridae) and groupers (Serranidae). These animals camouflage themselves amid the rocks by virtue of their markings and dash out to snatch prey as it passes. Eels prove particularly successful among rocks and wrecks in many parts, and they may grow to great sizes, like the morays (Muraenidae)

This photograph of the European Sea-urchin *(Echinus esculentus)* clearly shows the extended tube-feet. At the end of each is a minute sucker which can adhere to the substratum and provide a grip for locomotion.

and the congers (Congridae). They often have fearful teeth with which they grasp their prey before swallowing it whole. By keeping their tails in their holes they can exert leverage to drag their struggling prey, usually reasonably sized, back to the security of their lair.

By their very nature, rocky bottoms are unsuitable for fishing in any way other than by line or trap. Bottoms of sand or shell gravel are generally far more important for commercial fishing, for not only do these bottoms lend themselves to fishing with trawl and seine nets, but the fish populations which occur there are frequently more useful to man. Among the sandy bottom fish are flat varieties like flounder (Bothidae) and sole (Soleidae) which actually spend almost all their adult lives on the seabed. Then there are those like cod (Gadidae) which feed on the bottom and live just above it. Most fish like these have larvae which spend a while in the plankton after hatching from the egg. Larval and juvenile flatfish look much like the contemporaries of their round-bodied relatives, but after they have

sunk to the sea floor a strange metamorphosis takes place and they become adapted to horizontal life. Some species like Plaice (*Pleuronectes platessa*) and flounder lie on their left sides, while others such as the Brill (*Scophthalmus rhombus*), Topknot (*Zeugopterus punctatus*) and Turbot (*Scophthalmus maximus*) lie on the right side. The eye from the underside migrates around the top of the head to the upperside giving the head a distorted appearance with two eyes close together.

The commercial fish of sandy waters feed to a great extent on the invertebrate fauna of the sand itself. Bristle worms, small crustaceans, brittle-stars like *Ophiura*, and other small fish species make up their diet. Present with these fish, which are technically known as bony fishes on account of their skeletons having relatively hard bones, are some of the cartilaginous fishes such as the sharks, skates and rays. Cartilaginous fish have, as their name implies skeletons made from unhardened cartilage and are considered to be of lower evolutionary standing than the bony fishes. Skates

Flatfish like this flounder may adjust their coloration so that they almost exactly resemble the substratum on which they lie.

and rays are particularly well modified for life on the seabed and they are flattened top to bottom, rather than side to side as are flat bony fish. They also have their mouths on the underside rather than at the front of the head. Many of the skates and rays feed by rooting about in the sand for small invertebrates such as gastropods, bivalves and bristle worms. Some are able to crush even the strongest marine snail shells because they have teeth which are highly modified for dealing with this sort of prey. Some species of ray have particularly interesting modifications to their body muscles. These modified muscles are able to generate considerable electric currents which the fish can discharge at will and use either to stun any prey which may

be within reach at the time, or alternatively to act as a type of underwater echo-sounding system which can help them find food, or their way, particularly in very cloudy turbid waters.

Many representatives of the predatory sharks will be met with in such seas throughout the world, although they are at their most abundant in tropical waters. A number of small sharks such as dogfish, live as scavengers feeding on dead and dying fish as well as invertebrates. Many of these along with the skates lay their eggs in horny sacs which are attached to weeds or other structures by means of tendrils. They are known in some parts of the world as mermaid's purses, and after the eggs have hatched they may occasionally be washed up on

Stingrays of exotic appearance may be found on the seabed. They tend to cover themselves with sand to hide from predators.

the shore. In addition to the flat fish and the skates, sharks and rays there will be many schooling fish in the shallow seas. Some of these live near the bottom, but others live in the mid and upper waters.

The sandy seabed will have a rich population of invertebrates, many of which, as we have seen, form an important part of the diet of commercial fishes. As with the sandy shore a great number of these will be found actually burrowing in the sand. When discussing shore bivalves we noted how they were adapted for life below the sand. Burrowing bivalves will abound if the nature of the sand and food supply are correct, but so also may some surface-living species such as the scallop. Scallops, of which many species occur throughout the world,

live free and unattached. They are able to swim upwards by repeatedly flapping their shells. The scallop is generally quite sensitive to the smell of an approaching predator such as the starfish and quickly takes avoiding action by swimming away. Other non-burrowers characteristic of sandy bottom surfaces are the cuttlefishes, relatives of the octopus. Cuttlefishes, like squid, have eight short suckered arms and two much longer ones which can be quickly protruded to snatch unsuspecting prey from a surprising distance. A pair of simple fins run almost round the oval body and slow swimming manoeuvres can be effected by their use. Like the octopus rapid jet propulsive movements can also be made. When alarmed, or seeking to escape from

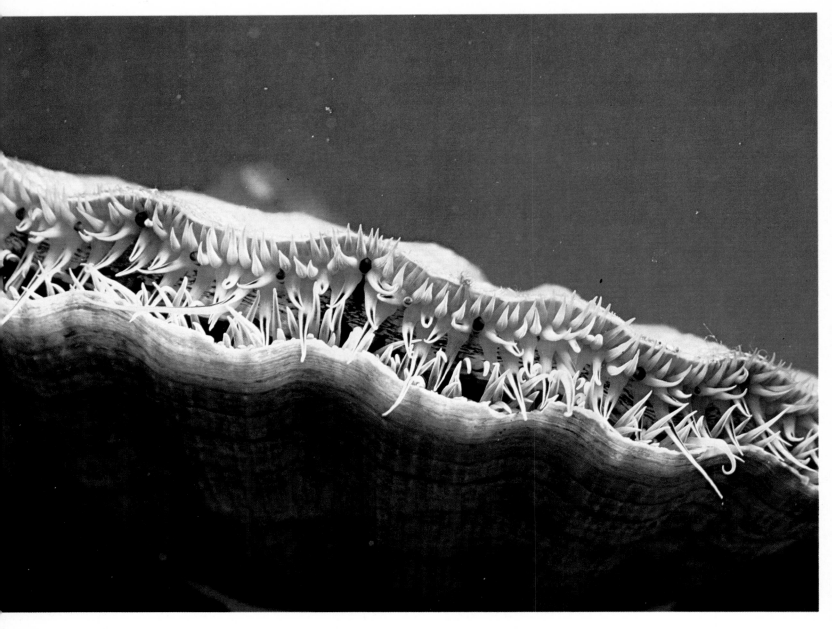

Above
The periphery of the scallop's *(Pecten)* body is fringed with sensory tentacles and eyes which quickly detect the approach of predators such as starfish and initiate the escape response of swimming.

Right
Seals are significant predators of fish populations in many waters of the world. In some areas they are regarded as serious pests by fishermen.

predators such as large fish, cuttle-fishes can squirt their ink into the water and make a dark cloud. This has two advantages, either it acts as a 'smoke-screen' behind which the cuttlefishes can retire, or it forms an alternative and distracting spectacle in the water upon which the predator concentrates its attention. Cuttlefishes also have amazing powers of camouflage and colour change.

A variety of crustaceans such as crabs, prawns, shrimps and their allies live on or in the sand, and in many parts of the world these animals, too, form the basis of an important commercial fishery. Brittle-stars, relatives of the star-fishes, are often present in great numbers, and indeed they may lie several individuals deep on the sand. How such a density of organisms receives its food supply is something of a mystery. In the main, brittle-stars live by scavenging bits of detritus and organic remains. Some species burrow while others

live on the surface and may form food for the commercially important species of fish.

The lack of any satisfactory anchor point will prevent the growth of most algae but so long as the water is clear enough, sea-grasses may flourish and the sand can be stabilized by their root systems so that other animals and plants can move in and live there. Large beds of sea-grasses grow in the Mediterranean, the Caribbean and along the east coasts of America. Some small algae may actually grow on the leaves of these grasses along with attached animals like hydroids and bryozoans. Such sea-grasses can form food for some marine mammals like the Dugong *(Dugong dugong)*, or sea-cows, or reptiles like turtles, which may be familiar in tropical and subtropical areas where the sea-grasses flourish.

Therefore, as on the shore, distinct and characteristic communities of animals and plants develop in the shallow seas.

Coral reefs

This aerial view illustrates the various reef zones. On the left the reef face slopes up from deep water. The reef crest is marked by breakers and the algal ridge and rubble zone lie behind. The back reef community slopes into the pale blue lagoon.

To many people the word coral conjures up images of white sand and palm trees on far away exotic islands. For those who have been lucky enough to have explored a coral reef, or gone skin diving on one, the first impressions will have probably been very different, but nevertheless unforgettable. Coral reefs provide some of the most beautiful and most complex of all marine communities. Because of the great diversity of species which we find in the tropics, an amazing variety of life-styles and fantastic body forms occur on them.

Although corals are but one of the many types of animals that are to be seen on a reef, they are in a sense the most important, for it is because they have developed there that all the other animals and plants have moved in too. Corals belong to one class of the phylum Coelenterata, the Anthozoa. Corals are carnivores and they trap small organisms using their tentacles. Characteristically they secrete a stony skeleton which surrounds the polyp like a cup, and into which the whole animal can withdraw. When it feeds the tentacles are protruded from the cup together with the upper part of the animal. The mechanism of skeleton secretion is the subject of considerable scientific investigation. Calcium and carbonate are extracted from the surrounding sea water and incorporated into the skeleton by a complex biochemical process. We have seen already that corals occur in all seas, but coral reefs are restricted to the tropics and are not formed by colder water species. Such reefs

do not develop in areas where the sea temperature falls below 18°C (64°F) and they only flourish where the temperature does not fall below 23°C (73°F). Even in the tropics not all corals build reefs. Those which do are described by scientists as hermatypic corals. Those which do not are known as ahermatypic corals. Hermatypic corals grow in colonies, that is, several or many coral polyps grow side by side secreting a common skeleton which can grow as a branching bush or tree-like structure like the coral genus *Acropora*, or as massive rounded colonies, as in the genus *Porites*. A large coral colony may include many thousands of living polyps all of which will resemble each other quite closely. In some species these polyps are relatively large and may measure several centimetres across, while in others they may be so small that a microscope is needed to make them out. Ahermatypic corals may be solitary like the cup-corals we have previously met in the shallow sea, or they may be arranged as small colonies. They seldom grow into large colonies, but some of the solitary ones may have enormous polyps as they do in the genus *Fungia*, the mushroom coral. Apart from the morphological differences between the two coral types there are also physiological differences. Hermatypic corals have minute single-celled plants called zooxanthellae living in their tissues as commensals. It is believed that their photosynthetic activities in the presence of sunlight are important in helping the formation of the coral's limestone skeleton. Since

ahermatypic corals do not need zooxanthellae they can form their skeletons in dark places such as caves or in murky or deep water where there is no light.

What is a coral reef?
A coral reef is a submarine ridge or platform formed by the accumulation of skeletons of dead hermatypic corals and other organisms associated with them in life. Over millions of years these accumulations of calcium carbonate have, in some places, amounted to very extensive concretions of chalky matter rising up many metres from the seabed. Reefs are constantly being added to by the skeletons of the latest, outermost generation of coral polyps. Consequently, they are continually growing. There are though, other forces which actively attack and break them down, consisting of the sea itself in the form of damage by waves and storms, and the animals which live amid the corals killing the polyps and drilling or penetrating the limestone.

There are three types of coral reef: fringing reefs which run along the edges of coastlines, barrier reefs which occur on the outer edges of continental shelves, and atolls which grow up from the ocean floor often capping undersea mountains. Hermatypic corals always grow towards the light and therefore towards the surface of the water. It has been suggested that a fall in the level of the seabed or a rise in the water level could have created circumstances under which coral reefs can develop vertically and horizontally to account for their present day forms.

Because of their temperature requirements, as mentioned earlier, hermatypic corals are restricted to tropical and subtropical latitudes. If one examines the globe in detail, reefs will be found to be well developed on the east coasts of continents but poorly represented on west coasts. The great oceans of the world are characterized by a clockwise circulation in the Northern Hemisphere and an anti-clockwise

current circulation in the Southern Hemisphere so that the west coasts of continents always receive colder waters from the poles, whereas the east coasts receive warmer tropical water favouring coral growth. Another interesting biogeographical point arises concerning corals. The waters of the tropical Atlantic are completely isolated from those of the Indo-Pacific, because of the cold water masses at the northern and southern points of the continents. These masses act as barriers against the passage of warm water species, preventing an extension of their range. About twenty-six coral genera represented by thirty-five species occur in the Atlantic. By contrast sixty genera are represented by several hundreds of species in the Indo-Pacific where coral diversity is vastly greater than it is in the Atlantic. Quite why this is so, is not understood, but it is probably related to the fact that the Atlantic is a relatively recent ocean formed as North and South America slowly separated further from Europe

Above left
In tidal lagoons massive corals, such as *Porites*, develop in characteristic micro-atoll forms with live polyps round the edges, which are always immersed, and dead limestone at the top where exposure to sun and air occurs.

Above
The branching coral *(Acropora)* develops in various ways. Here the stagshorn variety provides good cover for small fish and invertebrates.

Right
Fungia, the mushroom coral, is free-living as an adult and is not important as a reef builder, neither is it colonial. The ridges on the upper side of its skeletal cup resemble the 'gills' on the underside of a mushroom.

The brain-coral, *(Meandrina)* gets its name from the convoluted patterns of the individual polyps which make the colony look like a mammalian brain.

and Asia (continental drift), and is still undergoing the slow process of colonization.

The precise character and appearance of a coral reef depend on the environmental pressures leading to its development. As on the shore and in the shallow sea, tides, currents, exposure, rainfall and sedimentation all play a part. Within the three categories of reef, fringing barrier and atoll, many variations will be encountered. It is not possible to examine all these in detail here,

but a closer look at the community structure of, for example, a fringing reef might show something of the characteristics we might expect to encounter. Reefs develop quite differently on the side exposed to the prevailing weather so that as we pass from the exposed side to the sheltered side, we shall encounter ecologically different zones.

Typically, the reef rises sharply from the seabed as one approaches from the seaward side and may have a cliff-like face of 30 metres (98

Soft corals such as this one have skeletal elements as free chalky spicules not fused together. Hence the whole structure is soft and pliable.

Overleaf
Linckia laevigata is a conspicuous starfish in the reef waters of the Indo-west Pacific area. It is seen here creeping over a colony of hermatypic corals whose green tentacles are extended in many individuals.

feet) in height, or more. As the reef rises from the seabed there may be one or two ledges or terraces at various depths. The bottom of the cliff face, in the deepest water, is marked by a rather poor hermatypic coral growth because of the relatively low level of light. Further up, the coral becomes richer, and both massive and branching forms occur. These branching growths may assume flat, tabular or bracket shapes, or they can be more bush-like. On the terraces, the relative

flatness of the terrain allows quite large assemblages of coral to develop and pillar-like masses may reach up to several metres high.

Nearer the top of the cliff face, the corals are under greater influence from tides and waves. Here wave action is greater and oxygen and food supplies are richer. This point marks the transfer from the reef face to the reef crest where the reef begins to level out. Delicate branching colonies would be smashed by the waves so smaller, stubby, rugged

Looking out from the shore one gets a good impression of the shallow sandy floor of the lagoon which is often populated with coral outcrops.

colonies now appear. When the tide recedes there will be a great volume of water cascading down from the flatter part behind the reef and this can wear grooves in the reef crest. In some reefs these grooves are very large indeed, 3–5 metres (10–16 feet) across and as many deep. The movement of water in them may be so great at certain states of the tides that only flat encrusting organisms can grow against their walls.

The upper point of the reef crest

may be just submerged at high water and it is marked by the start of the third zone known as the algal ridge. The algal ridge is normally the highest part of the reef and is the area exposed to sun and air for the longest period and which bears the full force of the breakers in bad weather. Few if any animals can survive in this rough sort of environment but certain plants can. The calcareous algae are conspicuous here. Such plants frequently well represented are

Lithothamnion and *Porolithon* which also live elsewhere on the reef and, of course, in the temperate zones of the world. Their tissues secrete hard chalky crystals which make the plant appear more like a coral than a typical alga.

From the algal ridge the reef surface slopes down a little and, walking towards the sheltered side of the reef, one might come next either to an extended region of short coral growth in the case of flat-topped reefs, or directly to the so-called rubble or boulder zone. The initial appearance of this zone contrasts strongly with that of the reef face and the reef crest where live corals were much in evidence. In the rubble area we will find millions of fragments of broken coral, some large and some small, almost all dead and abraded by the action of waves and wind. These coral fragments originated mainly on the reef crest and parts of the reef top, especially in the case of reefs with an extended plateau-like surface. Rough weather causes them to break off and be washed over the algal ridge in a continuing process of breakdown. The rubble has a distinct flora and fauna of its own, based on the photosynthesis of encrusting and filamentous algae which grow on the dead coral fragments. The myriads of crevices and nooks among the rubble provide shelter and living space for a variety of fishes and invertebrates. In the case of fringing reefs a lagoon often separates the rubble zone from the sea shore and its sandy bottom is to some extent generated by the wave action grinding up the coral rubble. Lagoons have peculiar physical conditions all of their own, and they may be very productive in terms of the living plants and animals that they support. Beds of sea-grasses often develop in such lagoons. In the case of atolls a somewhat similar lagoon may be present within the circular shaped reef but on other offshore islands, different zonation patterns may have developed.

Animals living on coral reefs

Having dealt with the formation of coral reefs themselves we must now turn our attention to the other

The fire-corals *(Millepora)* have powerful stinging cells in each of their many defensive polyps. These are grouped around individual feeding polyps. The stings easily penetrate the human skin, hence the name.

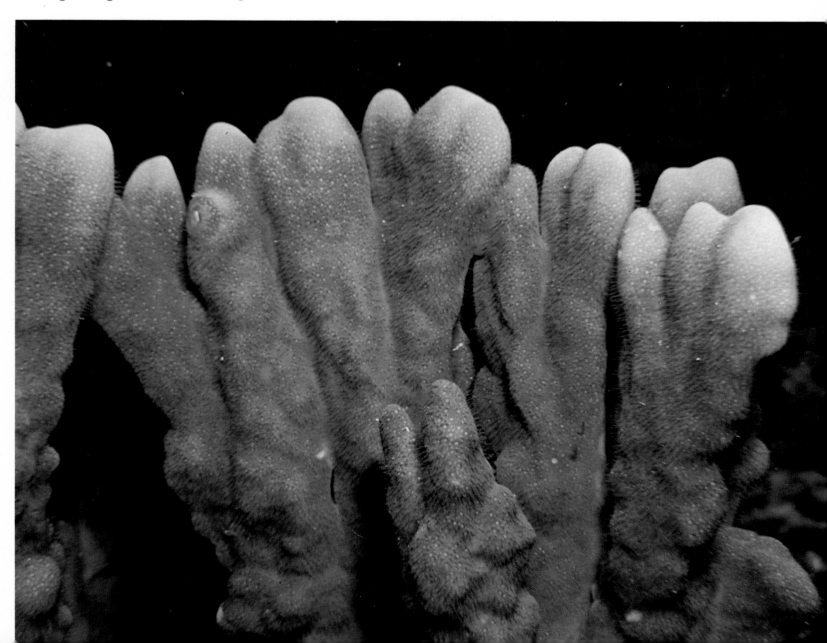

Right
The long spines of *Diadema* are not an absolute defence against predators. Several types of fish, including triggers and puffers, will attack it for food as will certain sea-snails.

Below
Cone shells are so-named because of their shape. The snails are predators and kill their prey by means of powerful venom. In the case of the Geography Cone (*Conus geographicus*) illustrated here the venom is deadly to man.

fantastically strong teeth bite off pieces of live coral. Some researchers think that these fish also attack dead coral to feed on the algae growing over the limestone.

A more bizarre form of coral predation is shown by the Donkey-fishes *(Bolbometropon muricatus)*, so named because of their size and coloration. These great fish have a pronounced hump on the top of their heads. They swim, or more accurately charge, into stands of branching corals and smash them down into smaller fragments which are then ingested for food.

The health of corals can be assailed in other ways. Disfigurement of coral skeleton growth can be achieved by animals such as the gall crabs. Here the female crab lives on a coral branch and causes it to secrete a capsule almost all round her. The male supplies her with food which he passes through a small aperture and she is able to get on with the task of reproduction well protected by the walls of her stony prison.

It is sad to reflect at this point that man has played a great part in the destruction of coral reef environments. A whole spectrum of interference from industry to

tourism is currently in evidence in many parts of the world. Not only are direct mechanical assaults responsible, such as occur in the construction of harbours, piers and warehouses, but secondary effects such as a change in the sedimentation rate of river-borne particles, due to dredging and other submarine works, can also be fatal. Then there is the problem of chemical pollution which has almost become universal. Finally, problems are sometimes created by tourists, reef-walkers and divers, who collect rare specimens like shells, some of which may take many years to reach reproductive maturity, and which therefore are vulnerable to over-collecting.

Other reef dwellers. The organisms which neither build nor destroy a reef display an immense wealth of diversity and only a few examples can be discussed here. In most cases the corals either serve as an indirect food supply or they provide the safety of shelter from predators. A great number of small crustaceans such as snapping shrimps *(Alpheus)* and small crabs conceal themselves in branching coral colonies. There they find protection from large predators. Snapping shrimps are quite remarkable. One of their two nippers is greatly enlarged and contains a powerful closer muscle. When the nipper is open the closer muscle is gradually tightened until a great tension has been generated. Then the nipper is allowed to snap shut creating a loud click that may be heard for some distance under water. Small prey organisms within a few milli-

Below
Groupers are territorial fish which feed on other fish and invertebrates. In this photograph a grouper is being cleaned by a cleaner wrasse, showing conspicuous blue and black markings.

Right
Much has been written about the association between anemones and fish. Anemone-fish, like the Clown-Fish *(Amphiprion frenatus)* shown here, appear to be protected from the host's stings by the slime it secretes.

Below right
Many of the branching corals afford shelter and protection for the numerous small reef fish like these damsel, or humbug fish.

This photograph illustrates clearly the normal and the inflated state of the Puffer-fish *(Diodon holocanthus).*

metres of the shrimps will be temporarily stunned by this and they are pounced upon before they recover. Another shrimp worthy of mention is the Cleaner Shrimp *(Stenopus hispidus).* This is larger and lives in crevices on the reef or in lagoons. The shrimp is clearly recognizable as it is banded brown and white. Larger fish which are infected with parasites on the scales and in the gills present themselves to the shrimp with a characteristic behaviour. The shrimp then climbs over them and removes the offend-

ing organism with its nippers. Not only crustaceans clean. A particular species of wrasse, the Cleaner Wrasse *(Labroides dimidiatus),* which is clearly marked blue with dark stripes, also sets up its 'surgery' on the reef at particular spots. When an infected customer presents itself the Cleaner Wrasse swims up to it, or even into its mouth, and bites off the parasite. This story is however more complicated because the Cleaner Wrasse has a mimic which exploits the behaviour of infected fishes. It is a species of

Right
The stripy markings and frilly fins of the lion-fish *(Pterois)* cause smaller fishes to mistake it for a plant or coral. As they seek shelter they are engulfed by its deadly jaws.

Below
An extreme modification for foraging for food in crevices characterizes the Long-nosed Butterfly Fish *(Forcipiger longirostris)*.

The Trigger-fish (*Rhinecanthus aculeatus*) gets its name from a dorsal spine which is normally lowered out of sight in a groove. When danger threatens this spine can be lifted, locked into position and used to wedge the Trigger into a crevice from which it cannot be pulled.

blenny, the Sabre-toothed Blenny (*Aspidonotus taeniatus*) which makes use of the cleaning station too. However, when the infested fish appears the blenny swims up to it, apparently with good intentions, but then it quickly bites a piece out of the fish and makes off with its meal! Phenomena like these make very interesting examples in the study of evolution, and fishes with amazing adaptations such as these will be found in many niches on a coral reef. Such a one is that exhibited by the clownfish genus *Amphiprion*.

Scattered around various localities on many Indo-Pacific reefs will be found large sea-anemones of the genus *Stoichactis*. These anemones may be discovered sheltering a number of brightly

coloured clownfish. The clownfish are protected from predators by the stinging cells of the anemone to which they themselves appear immune. In return they probably bring scraps of food to the anemones. The mechanism of the clownfish's immunity has fascinated scientists for some time, and it seems likely that a clownfish must get its body coated with anemone slime before it can pass safely through the tentacles of its host.

If one swims around a coral reef with a mask and snorkel something of the range of fish types should be apparent. Small damsel fishes (Pomacentridae) of various colours and patterns swim in clouds near the coral colonies. They forage for food in the convection columns of water which rise up at the reef's

edge. Large territorial groupers like *Epinephelus merra* will be seen lurking in crevices, as will the moray eels *(Echidna* and *Gymnothorax)* with their ferocious jaws. Both these types of fish live in a specific place, but the groupers emerge to forage on other fish and invertebrates, while the eels wait to grab at passing prey, only occasionally leaving their holes at night. Also to be found under the shade of coral overhangs will be porcupine and puffer fish. The latter can inflate their bodies with water if danger threatens and make themselves appear very much larger. The Porcupine Fish *(Tragulichthys jaculiferus)* and its relative the Box Fish *(Rhynchostracion nasus)* secrete strong poisons into the water which can have a serious effect on would-be predators. Box Fish, as their name implies, are completely encased in bone, and can only move their fins, jaws, gills and eyes. Brightly coloured butterfly fish (Chaetodontidae) with long snouts swim among the coral colonies. Different species have different lengths of snout so that they can feed specifically on small organisms at different depths in crevices and among coral

colonies. Then there are the herbivorous fish which graze algae, instead of eating other animals. The surgeon fishes (Acanthuridae) are one example. They get their name from two small movable projections looking like surgeons' scalpels which can be extended on the tail stalk. These projections are very sharp and can seriously cut the hand.

Apart from all the fish which actually live on the reef there are the visitors from the more open waters. Carynx, tunnies, barracuda and sail fish are but a few types. Then there are the various sharks which hunt for ailing and injured fish and which may be accompanied by the Pilot Fish *(Naucrates ductor)* or by Remoras *(Remora remora)*. Remoras are very interesting because their dorsal fins are modified to act as a sucker. In this way they can attach themselves to the underside of a shark, or other large fish — it has even happened to a diver — and hitch a lift. It is to be hoped that these very brief descriptions will convey something of the immense range of body forms and life styles which can be encountered on a coral reef.

A shark cruises just below the surface in search of prey.

Open seas

A variety of habitats is provided for many fascinating organisms, some extraordinary in appearance and mysterious in habit, by the vast expanses of the ocean. Much of what will be said here about surface waters applies equally to the shallow seas or neritic provinces. There is no sharp boundary between the surface waters overlying the continental shelves and the waters of the oceanic provinces, because various currents cause mixing to take place. Nevertheless, where there is minimum interference with salinity and temperature, the ocean becomes the most stable of all the marine environments.

This chapter will be concerned with pelagic life (that is, life in the sea's upper layers) in the top 150 metres (490 feet) of water. In this zone, two broadly different types of life can be found. On the one hand there are those organisms which spend their entire life cycle in the water and on the other there are those which spend most of their lives on the seabed, but also partly afloat.

For these two types of floating organism, certain basic problems have to be overcome. The sea must provide oxygen and food, as well as removing carbon dioxide and waste substances. It is important for pelagic animals to be naturally buoyant too, otherwise they would have to maintain their mid-water position by swimming movements which would be costly in terms of use of food and energy. Although drifting floaters or plankton (typical of many of the pelagic organisms we shall encounter) may use swimming

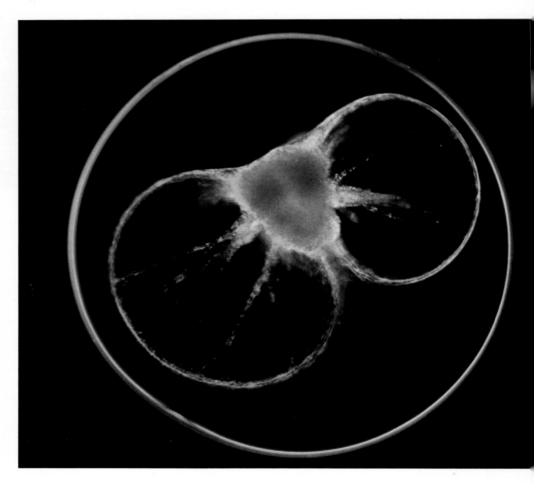

movements to keep their vertical position, they rely chiefly on their natural buoyancy. The way this is achieved is rather interesting.

Various means of floating are used by pelagic animals, but we will deal with the principal two. First, by evolution an organism may have increased its surface area/volume ratio so that, for a relatively small volume, it has a large surface area. Such an organism, while still tending to be heavier than the surrounding water, will float more easily because the surface area consists

This is a photograph of a dinoflagellate, an important microscopic constituent of the phytoplankton. Some dinoflagellates produce brilliant phosphorescent-like showers of sparks in the water at night.

of fin-like projections or bristles which can be picked up by water currents which then waft the organism along. When it sinks, it will sink more slowly than if it was of a different shape.

Second, the organism can reduce its absolute density, thus making itself lighter than the sea water which supports it. It can, for instance, develop fat bodies or oil globules in its body cells, or even remove heavier metallic components of its tissues and replace them by lighter ones, like ammonia. The organism may form a raft of gas bubbles outside its body, or it may have a swim-bladder functioning like a gas bag inside its body. This last has the advantage, as with the bony fishes, that the quantity of gas can be regulated to control the degree of buoyancy.

Nekton is a term applied to actively swimming organisms, as opposed to drifters. In practice it is generally associated with fishes, the most significant and important swimmers, but this has not always been the case. At one time in the Earth's history, the squids ruled the seas and even today they are conspicuous members of the nekton. Most of the smaller swimming invertebrates such as molluscs, worms and some of the arthropods are really classified as plankton because they cannot swim powerfully enough to counter the currents.

Earlier, the question of food chains arose. As we have seen, on the shore and on the coral reef, plants are the primary producers and upon them the energy supplies of the herbivores (plant-eaters) and the carnivores (flesh-eaters) ultimately depend. In the chapter on shallow seas we saw how the productivity of phytoplankton (plant plankton) is governed by the seasons which set the rate of photosynthesis according to temperature and daylight length, and how the availability of mineral salts in surface waters is largely governed by the thermocline, itself a seasonal characteristic. The relationships between plant and grazer, prey and predator, are as complex in the plankton as elsewhere, but without an adequate supply of plant material, the animals cannot flourish. Thus, the bursts and declines of the phytoplankton are quickly reflected by bursts and declines in the zooplankton (animal plankton).

Many types of phytoplankton are known, but the two major groups,

already mentioned, are dinoflagellates and diatoms. These are both unicellular algae, although the diatoms may occur in rod or chain-like colonies of many single cells grouped together.

Dinoflagellates are characteristically equipped with long whip-like threads called flagella. These can be lashed in a sinuous manner by the cell to propel it through the water. At the base of the flagellum, just inside the cell, there is a pigment spot associated with a light receptor. The effect of light falling on the pigment spot and the receptor is to cause the flagellum to beat so as to drive the cell towards the source of the light. In this way the dinoflagellate ensures that it keeps near the surface of the sea water under conditions ideal for photosynthesis.

Unlike the dinoflagellates, the diatoms have no motile flagellar processes. They depend for their flotation on the complex fin or

A plankton haul viewed under the microscope shows a variety of forms of marine diatoms.

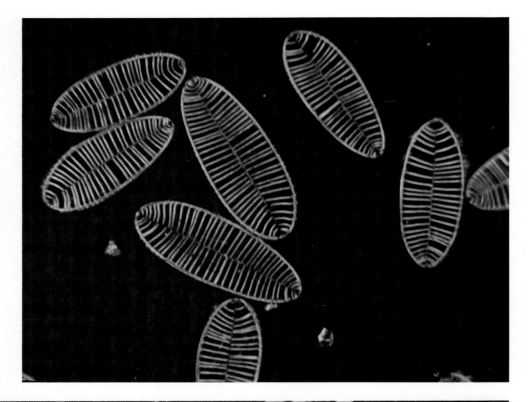

bristle-like extensions of the cell wall, and upon oil droplets contained within the cells. They are wafted around by the water currents which keep them in the photosynthetic zones for considerable periods. They absorb oxygen for respiration, carbon dioxide and water for photosynthesis, and mineral salts for protein elaboration directly from the surrounding sea water in just the same way as do the dinoflagellates.

Carefully controlled sampling programmes carried out using very fine meshed plankton nets have shown that during the hours of daylight the phytoplankton tend to move near to the surface of the sea. This is the reverse of the twenty-four hour cycle as shown by most zooplankton, which rise in the water to feed near the surface at night and sink to the deeper water by day, where they are probably less conspicuous to predators which hunt by eyesight. The floating microscopic algae are normally invisible to the naked eye but there are occasions when, under very favourable circumstances, so many millions can be aggregated together that they become visible in the form of blooms, or windrows, which colour the surface of the sea reddish-brown. This colour is due to the pigments of the individual plant cells gathered together. Under these circumstances they form the so-called red tides which are a feature of certain points around the Central American coasts and other parts of the world like the Red Sea.

Not all the floating algae are microscopic. One well-known example that is not, is Sargassum which, in a number of species has both a floating and an attached stage to its life cycle. Sargassum weed gives its name to the Sargasso Sea, and the moderately large floating plants which accumulate there harbour a fauna of amazingly modified animals, like the fishes with fin shapes and skin coloration which make them almost unidentifiable amid the weed fronds.

The first category of animal mentioned at the start of this chapter was that consisting of those organisms which spend the whole of their lives afloat. With the exception of those sedentary phyla like the sponges and bryozoans, and one or two groups of worms, almost every major animal group has full-time plankton members. This is even true of the sluggish, heavy echinoderms which have floating representatives like the sea-cucumber *Pelagothuria*. Such animals can be classified as herbivores, omnivores and carnivores. Some are filter-feeders while others have evolved peculiar yet efficient ways of hunting specific prey.

The coelenterates have several fully planktonic representatives. Well known for their powerful stinging cells are the siphonophores, of which the Portuguese-Man-o'-war (*Physalia physalis*), is an example. *Physalia* is really a colony with various types of polyps modified to fulfill particular roles. One polyp acts as a float, bearing an air sac, others are concerned with feeding, reproduction and defence. Long trailing polyps armed with powerful stinging cells capture the prey as *Physalia* drifts along. The prey consists of fishes and other moderately sized animals. A near relative of *Physalia* is *Velella* which is a colonial coelenterate belonging to the chondrophores. Its common name, By-the-wind Sailor, indicates the method of propulsion: on top of the animal is a crescent-shaped sail of tissue which catches the wind.

The phylum Ctenophora is closely related to the coelenterates. Ctenophores are known as comb jellies. They do not have stinging cells but they have specialized cells called colloblasts. These may be trailed behind on retractable tentacles like microscopic fishing lines. This is the case in *Pleurobrachia pileus*. Small organisms that collide with them are

Above left
At high magnification the disposition of green photosynthetic chlorophyll pigment can be clearly seen within the diatom's body.

Left
This photograph illustrates the flotation bladders of Sargassum weed. It also shows an anemone and a Sargassum fish with plant-like fins.

Above right
Some of the long trailing tentacles of the Portuguese-man-o-war (*Physalia*) and the inflated air sac are shown here. Many tentacles are retracted. These animals occasionally get driven on to the shore, when the tentacles are often broken off leaving the air sac alone stranded.

Right
The chondrophore By-the-wind Sailor (*Velella*) has shorter tentacles than *Physalia*. The darker area within the disc marks the extent of the horny skeleton which supports the body and the sail.

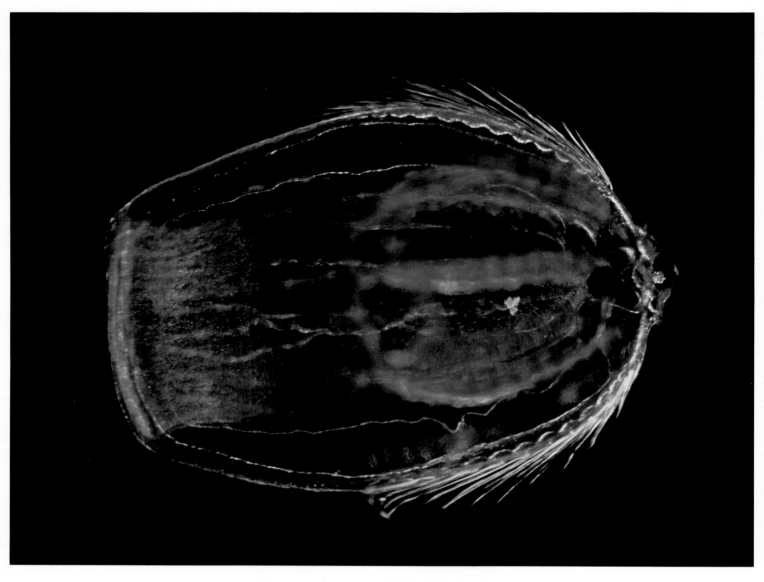

trapped by the lassoes of the collo-blasts and the food is then wound in and passed to the mouth. Cteno-phores are important planktonic predators, and feed extensively on small crustaceans and other food some of which is suitable for com-mercially important fish. They swim by means of special organs called comb-rows which are arranged in parallel rows up and down the body. Each row has many fine projections grouped together like the teeth of a comb. When the combs vibrate in sequence the animal is propelled through the water. A simple balance organ helps to orientate them in the water.

There is a variety of fully pelagic animals within the phylum Mollusca. The snail *Ianthina exigua*, for example, floats on a raft of bubbles specially secreted for that purpose. *Ianthina* feeds on siphonophores and chon-drophores, apparently undeterred by their stinging cells. Another group of sea-snails which show a far greater modification to pelagic life are the pteropods or sea-butterflies such as those belonging to the genus *Clione* or *Limacina*. Pteropods, mol-

luscs with the middle part of the foot expanded into wing-like lobes in a pair, are most common in tropical waters but also occur in the temperate seas. They are able to swim by means of paddling their wing-like lobes. *Limacina*, in fact, appears to row itself along. They may also have tentacles and other appendages. There are two sorts, one with and one without a shell. For food they rely mainly on dino-flagellates and diatoms but some are carnivores, feeding for example, on other species of pteropod. Some pteropods also form an important food source for pelagic fish, such as herring. These molluscs lay their eggs in floating strips of jelly.

The squids, as we have already seen, are important members of the pelagic fauna. Some species are minute while others are enormous. They display highly developed sen-sory organs and sophisticated move-ment patterns which allow for care-fully controlled manoeuvres, includ-ing jet propulsion to escape from predators.

Certain whales prey upon the giant squids, and it is not un-

common to find the impressions of squid's suckers on the bodies of captured whales, a rather grim testimony to the underwater battles that probably take place before a squid is engulfed. In fact, some squids are known to science only through their remains having been found in the stomachs of whales.

A few annelid worms are fully planktonic. The transparent worms of the genus *Tomopteris* are an example. They have well developed paddle-like structures arranged in series down both sides of their bodies. By wriggling in a sinuous manner they can swim rapidly through the water. Being almost completely transparent they are well protected from predators which hunt by sight. Indeed, when cap-tured in a plankton sample their presence may only be detected by the swirling of the material sus-pended in the water.

The arthropods are well represented in the plankton by the crustaceans which swim or float.

The copepods (a subclass in the great phylum Arthropoda) are principally a planktonic group and

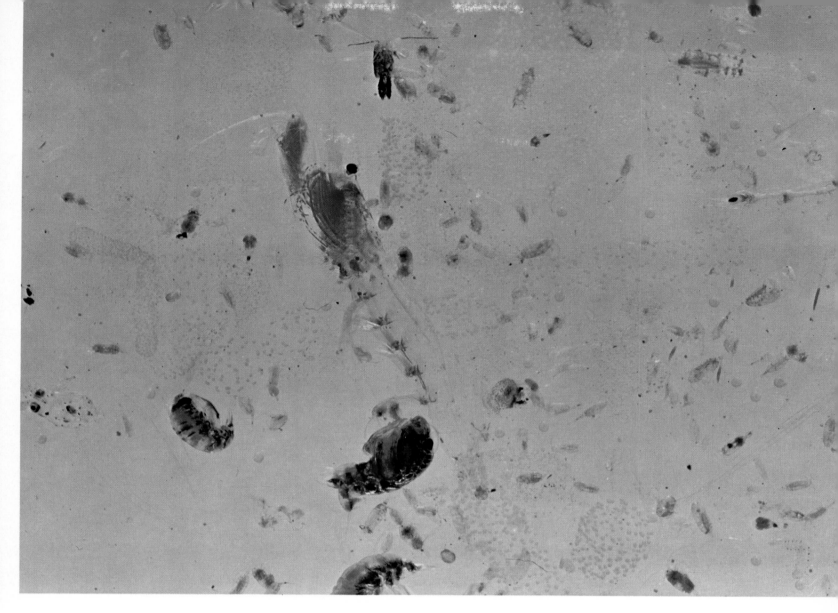

they contain some of the most important herbivores. Copepods are small crustaceans which swim by lashing the water with their oar-like antennae. By doing so they lift themselves in the water and they create complex mini-currents in the region of their bodies. These currents bring minute food particles such as phytoplankton within reach of their other appendages which then filter them from the surrounding water and pass them to the mouth. Copepods are in turn preyed upon by a variety of fish and other invertebrates and they are of great economic importance to fisheries. One particular copepod *(Calanus)*, is, in fact, the most important herbivore of the seas of the Northern Hemisphere, while in the Southern Hemisphere the foremost crustacean is the larger, more shrimp-like, *Euphausia*. *Euphausia* is so common in the Antarctic that it forms the food of the Blue Whale *(Balaenoptera musculus)* and is known as krill.

Space does not allow a full description of all the other remark-able crustaceans that may be found in a single plankton sample, but a few must be mentioned. The Glass Shrimp *(Phronima sedentaria)*, which is not really a shrimp at all but a relative of the sand hoppers (members of the Order Amphipoda), has huge eyes and a body compressed sideways. It lives inside a weird barrel-shaped 'house' made of glassy jelly. By beating its abdominal appendages it pumps water through the barrel which is open at both ends, and so it scuds about in a sort of jet-like fashion. The origin of the barrel is interesting. It has been stated many times that the barrel comes from the body of a salp (see below), and that the Glass Shrimp enters the salp and eats out its body, thus leaving the empty barrel case. Other scientists believe that since the shrimp has not been captured without the 'house', that it is probably made by the shrimp itself.

Stalked or goose barnacles are one group that may colonize tar-balls or other floating objects such as driftwood, and electric light bulbs for that matter. The genus *Lepas* includes many species of goose

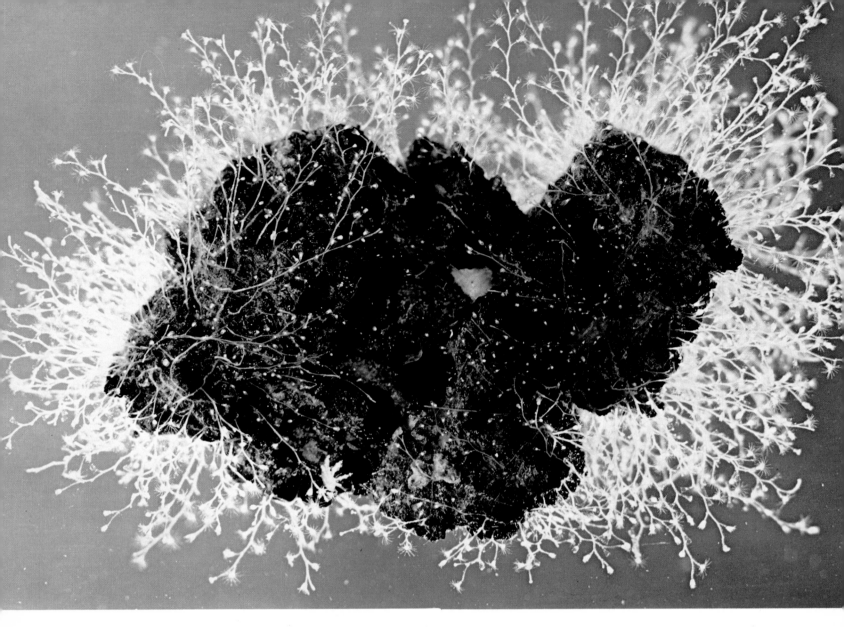

Hydroids as well as stalked barnacles will find room to settle and develop on tar-balls such as this. Here some fine hydroid colonies can be seen.

barnacles. They feed with large filtering baskets and take in a variety of planktonic organisms. There are even a few species which need not acquire a float since they are able to develop their own. These types are known as buoy-making barnacles. Tar-balls also support other crustaceans such as the flat isopods, *Idotea*. Species of *Idotea* chiefly occur on the shore and in the shallow sea but have also evolved to live on tar-balls, where, presumably, they feed on bacteria. Their bodies are coloured with dark pigment so that they are relatively inconspicuous.

The phylum Chaetognatha is almost exclusively planktonic. Chaetognaths are small predacious worms much flattened from side to side. Because of their general shape and the way they dart through the water in quick bursts, they have become known as arrow worms. At the head end there are well developed jaw-like structures which seize and swallow the prey. Chaetognaths are sensitive to vibrations in the sea water, such as those caused by the swimming movements of their prey, which include crustaceans. When stimulated by such vibrations the arrow worms home onto the prey and snatch it. They, along with the ctenophores, are ecologically important planktonic predators.

The salps have already been mentioned. They are planktonic relatives of the sea-squirts, familiar on the seashore and in the shallow seas. They and the sea-squirts are known as tunicates, because they have a special outer tunic. The salps have very complicated life cycles. As adults they swim through the water using ciliary power from their gills. At one end of the barrel-shaped body is the inhalent opening, and at the other end is the exhalant one. As water is pumped through from one end to the other, food particles, principally phytoplankton, are removed, as they are in the sea-squirts. Rings of muscle running around the inside of the barrel or tunic confer manoeuvrability. For reproductive purposes many salps trail a streamer-like stolon through

Goose barnacles *(Lepas anatifera)* make use of floating objects such as driftwood or even electric light bulbs to support them in the open sea. In some of these specimens, the basket-like arrangement of thoracic appendages can be seen extended for feeding.

the water with young organisms developing on it. Some salps are solitary, whereas others can form colonies up to 50 centimetres (1·6 feet) in length. Many are phosphorescent and if taken at night can flash and glow spectacularly.

In all parts of the world there are fishes that live entirely in the surface waters both as larvae and as adults. In temperate seas herring-like fishes may swim in vast shoals feeding on copepods and other small crustaceans in the zooplankton which they filter from the water by means of rakers or filter bars on their gills. Relatives of the herrings which occur in the tropics have highly developed fins and are known as flying fish. Some species just have the pectoral fins modified for gliding, but in others the pelvic fins are also adapted. It is now known that these fish cannot actually accelerate by using their fins in air. The fins act like the wings of a glider and the fish pick up power under water and leap into the air in order to glide. By lashing their tails in the surface layers they can maintain their air-

borne progress for some little distance. Flying fish probably take to the air to avoid predators such as tuna, but in doing so they expose themselves to the attacks of predatory seabirds such as gulls and albatrosses.

The tuna-fishes (Thunnidae) are another group of surface dwellers. Some species of tuna reach a considerable size, for example, the Mediterranean Tuna *(Thunnus thynnus)* is said to reach over 2 metres (7 feet) in length. Tunas are predators and when small feed on crustaceans, but as they grow they take small- and moderate-sized surface fishes such as sprats and herrings.

There are a great number of different species in the tuna family. Some, such as the Mackerel *(Scomber scombrus)*, are migratory and spend more time in the colder waters during the summer months. While the tuna family shows remarkable bodily modifications for fast swimming, in that they can tuck their fins out of the way into grooves, they are not as a group as highly

79

modified as some of their co-
residents in the surface waters. The
Gar-fish *(Belone belone)* has an
astonishingly elongated snout which
is used for selecting and snatching
smaller prey; then there are the
half-beaks which only have one
jaw extended. Barracuda *(Sphyraena)*
are among the most voracious
pelagic fishes in tropical waters
where they may reach 2 metres
(7 feet) in length. Like the sail-fish
(Istiophorus) and some of the tunas
they are highly regarded by sports
fishermen. Sail-fish, allied to the
swordfish, have a highly developed
dorsal fin which is thought to be
used not for fast swimming, though
they can move very rapidly, but for
rounding up small prey. By raising
the fin the fish presents a large
surface area when viewed sideways and by beating the water
with its highly extended upper jaw

it can herd small fishes and eat
them. When swimming quickly the
dorsal fin is lowered into a groove
on the fish's back.

Another remarkable surface fish
is the ocean sunfish *(Mola)*. This
extraordinary animal looks to be
mostly head. When viewed from
the side it is almost circular. It
swims by waving narrow but extended dorsal and anal fins, the tail
acting as a rudder. Ocean sunfish
feed on jellyfish and other plankton
and can reach an immense size.
One species may reach about 2
metres (7 feet) in length and weigh
a tonne or more.

No account of life in surface
waters would be complete without
reference to the great whales. Blue
Whales are the largest living organisms, and some weigh up to thirty
times more than an elephant.
Whales are mammals, and breathe

air like humans. Some species such as the Sperm Whale *(Physeter catodon)* are carnivorous and feed on fishes, or squid and occasionally seals. Others, of which the Blue Whale is an example, are planktonic feeders and use the extraordinary whalebone filters in their mouths to separate food from the water which they suck in. Krill forms a very important part of the diet of the whales and one large Blue Whale may filter as much as 10 tonnes of krill from the sea in order to increase its weight by one tonne.

Sadly, the nations of the world are unable to agree on a policy for the conservation of some whale species, and in view of the sophisticated methods of whale hunting which now employ helicopters and high-speed vessels, some species of these great mammals are con-

sidered by many authorities to be under threat of extinction.

This account of life in the surface waters of the open sea has so far neglected what might be called the part-time members of that community. These organisms comprise, to a large extent the countless eggs and larvae of those organisms which as adults live on the seabed. For a totally sedentary animal such as a sponge or a bryozoan, which as an adult is immobile, it is essential to have some form of dispersive stage in its life which will enable it to colonize new environments. A floating planktonic larval stage fulfils this need. Such larval development has another advantage, in that it means that the juveniles of bottom-dwelling species do not compete with their parents for food or living space.

Very often larval life is totally different from that of the adult

A school of dolphins breaks the surface. Under water the dolphins, like the great whales, appear able to communicate with each other by a sophisticated sound system.

of the same species. For instance, the adult starfish is a slow moving heavy-bodied carnivore which preys voraciously on any other animal it can get hold of. Its larva, on the other hand is a minute, floating animal of great beauty which feeds on phytoplankton. The adult has a peculiar symmetry with the mouth below and the anus above, as well as a five-sided, somewhat rounded body. The larva, however, has a clear left and a clear right side which its parents lack as adults. Space does not allow a fuller investigation

of larval forms, which are typical of all groups from the sponges to the fish, but something of the diversity of these important plankton components will be appreciated from the photographic illustrations provided in this chapter.

The communities of the surface waters of the open seas are, as has been shown, as diverse as any in the marine environment. Further, it is worth noting the importance of the the surface waters themselves as nurseries and means of distribution for the bottom-dwelling organisms.

84

Deep seas

A large proportion of the Earth's surface is taken up by the great oceans. As we have seen already, the land masses are surrounded by continental shelves which stretch away from their shores in gentle underwater slopes. In some parts of the world, the continental shelves are very narrow, elsewhere they can be extensive.

At their outermost limits these shelves generally descend to about 150 metres (492 feet) below the surface. Here there is usually a change of topography. Beyond the 150-metre mark, the seabed normally begins to form a sharper downward incline. This change of angle in the slope of the shelves marks the transition from shallow seas (neritic provinces) to the ocean proper. The downward slope is technically known as the continental slope.

The continental slope may extend downwards for a further 160 metres (525 feet) and it ends when the level, plain-like floor, or abyssal plain, is reached. The depth of water from the abyssal plain up to the surface may be in the region of 3,000 metres (9,850 feet). The ocean bed stretches across vast distances until the surrounding continental slopes begin to rise up towards the land again.

Usually ocean floors are divided up into a number of basins by mountain-like ridges. (Elevations and depressions on ocean floors, as on land masses, are referred to as forming the relief structures of a given area.) As well as being mountainous, relief on the ocean floors may take the form of furrows or trenches.

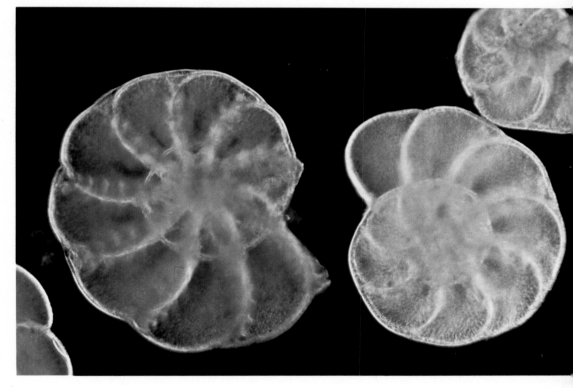

The deepest trenches on Earth occur in the Pacific Ocean, where the ocean bed lies about 10,000 metres (32,800 feet) below the surface in the Philippine Trench and 10,500 metres (34,450 feet) in the Japan Trench.

Early marine biologists like the Englishman, Edward Forbes, in 1859, divided up the living species in the ocean according to the principal organisms that occurred there and decided that life did not exist below 550 metres (1,800 feet). Beyond this limit, they said, was the azoic zone, or zone without a trace of life. In the 1860's expeditions were organized in Britain under the leadership of Dr Wyville Thomson. These determined that in the Atlantic

Foraminiferan skeletons of chalky material are responsible for extensive deposits on the bed of the deep sea. This is a view of skeletons under the microscope.

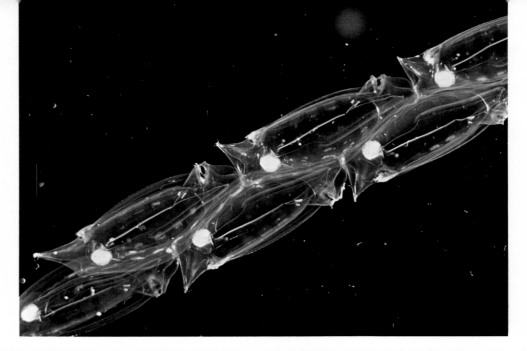

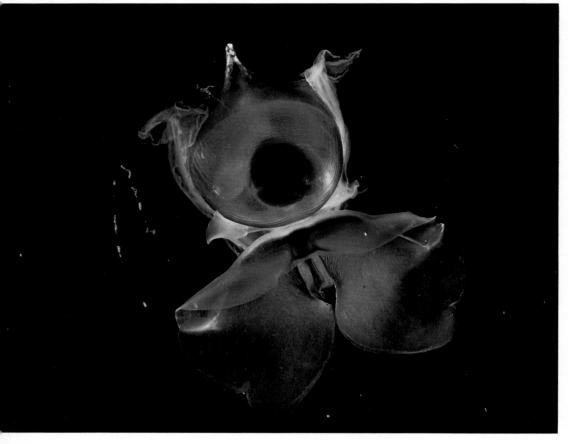

tained by several American ships in the Pacific during the late nineteenth century confirmed these findings.

As a result of improvements which were made to collecting equipment, it soon became apparent that a characteristic fauna occurs at each particular level of the ocean. Beneath the upper pelagic zone lies an extensive region of water known as the bathypelagic or mid-water zone. Research has shown that a considerable range of animals inhabit the various levels of the zone and that below it live the true bottom-dwelling creatures known as the benthos. If one had been able to journey down the continental slope, one would have noted that the typical shallow sea is quickly left and a characteristic continental slope community develops. At levels below 1,000 metres (3,300 feet) the true abyssal fauna appears and extends out on to the abyssal plain. So the two distinct faunas of the deep ocean are the bathypelagic fauna and the benthic fauna.

Before we consider the organisms themselves, we must pause to look at some of the physical conditions which they have to master. The deeper into the ocean one investigates, the greater becomes the pressure due to the weight of the water itself. At 1,000 metres (3,280 feet) depth the pressure is immense being about 1,000 times that of atmospheric pressure. When deep-sea organisms are lifted to the surface artificially the most profound changes occur within their bodies which more often than not either kill them or seriously damage them. In practice few animals move far vertically up or down, but if they do they must have some means of compensating for the changes in pressure that will occur.

When the water is clear and free from suspended matter, light can penetrate quite a long way. In ideal circumstances, light detectable to the eye may reach down as far as 500 metres (1,650 feet) and experiments with sensitive cameras have shown that it may reach further, but such light would not be much use to photosynthetic organisms, which are limited to the surface layers of the sea. Below 1,000 metres there is no daylight at all, so the greater parts of the ocean depths are in a state of permanent darkness and animals make do without light, or generate it themselves.

Top
Iasis is a colonial salp with many individuals strung together in a chain. Each individual is a functional animal with inhalant and exhalant siphons and a filter feeding pharynx. The nervous and sensory elements are greatly reduced.

Above
Cavolinia tridentata, a pteropod, has a characteristically pointed shell and a large foot divided into two conspicuous lobes or paddles which it uses to propel itself through the water.

Right
Brachiopods have their bodies entirely enclosed by two shells somewhat reminiscent of the bivalve molluscs. They also live on the seabed and filter-feed. Here one is seen attached to some dead deep-water coral.

an abundant and characteristic fauna existed down to about 2,000 metres (6,500 feet). Wyville Thomson's work, together with that of his able colleagues, stimulated the formation of the famous 'Challenger' expedition. The 'Challenger' sailed in 1872 with a distinguished crew of scientists and, after a round-the-world cruise returned to Britain in 1876 bearing an immense store of preserved specimens and much valuable data about life in the ocean depths. By then it was accepted that organisms existed at every depth and that the inhabitants of the abyss were highly evolved to suit the rigours of their deep habitats. Similar results ob-

Euphausia superba shows the typical krill body form. There are many fine thoracic appendages used for straining food from the water.

Under the influence of warm summers and cold winters the surface waters of the open sea will fluctuate in temperature, but the deeper waters below do so very little. At 2,000 metres the temperature is usually between 0° and 4°C. Similarly, changes in salinity occur at the surface due to evaporation and rainfall, but in water of any depth the salinity is very constant and from 2,000 metres down to 8,000 metres of between 34·5 and 35 parts salt per 1,000 parts water have been recorded.

What are the types of animal to be encountered in the depths of the ocean? First we will discuss the mid-water or bathypelagic organisms. A great proportion of scientific catches show that the invertebrates are quite similar in body form and life style to those found at the surface. For the greater part we shall be looking at a set of deep-water relatives of the surface organisms discussed in the chapter on the open sea. However, among these, especially where the fishes are concerned, will be others that have no counterparts at the surface.

Herbivores like copepods, pteropods and salps, as well as a multitude of invertebrate larvae, feed on the

phytoplankton. Of course not all the phytoplankton are eaten near the surface, and many live on to old age whereupon they die and gradually sink below the photosynthetic zones. As they do so they may be consumed by deep-water herbivores. It has been demonstrated scientifically that most of this plant material is consumed before it reaches a depth of 300 metres (1,000 feet). Copepods live in great numbers in the bathypelagic zones, and, as they swim through the water they extract some of this particulate matter. Although many of them are colourless and transparent, some of the deep water copepods are highly coloured, either by special pigments in their bodies or by the incorporation of oil and fat reserves. Salps too are important deeper grazers. Below 500 metres there will be little plant originating material for food, and instead the filter-feeders will have to concentrate on detritus and larger fragments of dead and dying animal matter.

Krill, crustaceans known as euphausids mentioned in the previous chapter, are significant members of the bathypelagic fauna. As we have already seen these animals are the principal food of many whales. They are small animals usually reaching from 1·5 to 3 centimetres (0·5 inches to 1 inch) in length. They have conspicuous eyes and quite powerful swimming appendages, and they seem capable of ceaseless activity. In the colder water species the appendages at the front of the body are modified for catching and handling particulate food such as phytoplankton and detritus. In the warmer tropical and subtropical seas it seems that such particles are not used and that instead these animals catch small fish, copepods and other planktonic animals. At similar depths we may also encounter large deep-red prawns, some of which may measure up to 12 centimetres (4·5 inches) in length or more. Some of these prawns may be strongly luminescent. It is thought that many of them feed on other crustaceans, small fish and animals like arrow worms. Others use their filter-like appendages to collect suspended detrital food from the water.

Many carnivorous invertebrates exist in the bathypelagic regions. Some like the chaetognaths (arrow worms) are very similar, though often larger than their surface-dwelling relatives. The coelenterates

Left
Parapronoe crustulum, an amphipod,
lives as a member of the deep-sea
plankton which forms the diet of a range
of animals.

Below
This photograph of the mid-water prawn
Pasiphaea illustrates the deep red
coloration characteristic of some
crustaceans from deeper waters.

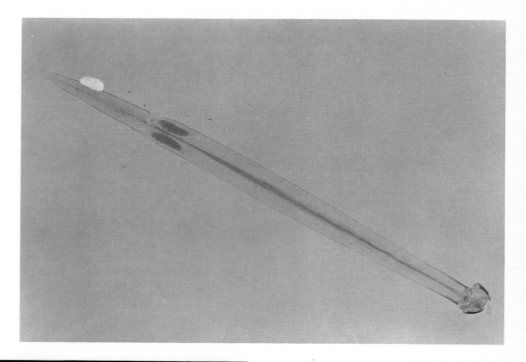

Right
Chaetognaths like *Eukrohnia fowleri* are found at various depths. This photograph shows the arrow-shaped body flattened from side to side, as well as the fins and head with voracious, bristle-like jaws.

Below
Deep-sea siphonophores like *Physophora hydrostatica* swim by means of their contractile bells. These can be discerned in the upper part of the picture. The prey-catching tentacles hang below.

are again represented by true jelly-fishes and siphonophores.

The true jellyfishes swim by repeatedly contracting their umbrella-shaped bodies. They trail behind them stinging tentacles which will trap prey from copepods to fishes.

The annelid worms and the molluscs will be represented in the bathypelagic zones by similar species to those found at the surface. However, in the case of the molluscs we may find some extra representatives. An example is that of the peculiar pelagic octopuses. Some species are beautifully modified for life in mid-water. If they live below the levels to which light penetrates they may have degenerate eyes, due to lack of use. Webs of thin skin link the tentacles together forming a bell which gives the animal a superficial resemblance to the jellyfishes.

No account of the bathypelagic community would be complete without reference to the fishes which inhabit the mid-water regions. By comparison with surface and shallow water fishes many of them are fantastic in appearance. Some live in zones where there is little light; others live in total darkness. Many, such as the lantern fishes and the hatchet fishes, so called because of their shape, have conspicuous luminescent organs on their flanks and bellies. Then there are the angler fishes *(Ceratiidae),* with the first ray of the dorsal fin modified to lure unsuspecting prey towards deadly jaws. Some of these animals are truly grotesque in appearance and are equipped with teeth which, in relation to the overall size of the fish are enormous. Often these teeth merely serve to prevent the escape of prey from the mouth. The lure may be augmented by luminous chin barbels. In some species of angler the male lives as a dwarf parasitic on the female. In this way the difficulty of re-finding a male each breeding season is overcome. Another weird deep-sea group is that of the gulper-eels (Eurypharynx). These long thin fish have an enormous head, entirely out of proportion to the diameter of their stringy bodies, and even more enormous jaws so that they can engulf prey of considerable size like other fish. The fish that live down to about the 1,000-metres level have well-developed eyes and optical centres to their brains. Those which dwell in the realms of perpetual darkness, have small or degenerate visual systems. Any eyes that may be found have probably become adapted to registering the presence of light emitted by the light producing organs of the other animals living with them.

Light organs occur in deep-sea fish, crustaceans and molluscs like squid. Other invertebrates may be luminescent over all or part of their bodies. There are two basic ways in which light can be produced. It may be developed biochemically in the presence of oxygen inside special glandular cells, or it may originate from cultures of luminiscent bacteria maintained by the host organism. Once the light has been generated it can be concentrated by special lens systems. Fish may use lights as a lure for prey and there are reports of some species which can concentrate strong beams for locating potential food. Another use is where particular patterns of light can aid recognition within and between species. There may be sexual differences in the patterning so that mates can be recognized. Finally, luminescence may be used to frighten off predators and to act as warning signals. In the case of sedentary bottom-dwelling invertebrates like the coelenterate sea-pens, a flash of light might well deter a predator.

Many of the principals of luminescence discussed above apply equally well to those animals which live permanently on the bottom of the ocean. Benthic animals are fundamentally

The rat-tailed fish (Bathygadus) has a long tapering body from which it gets its name.

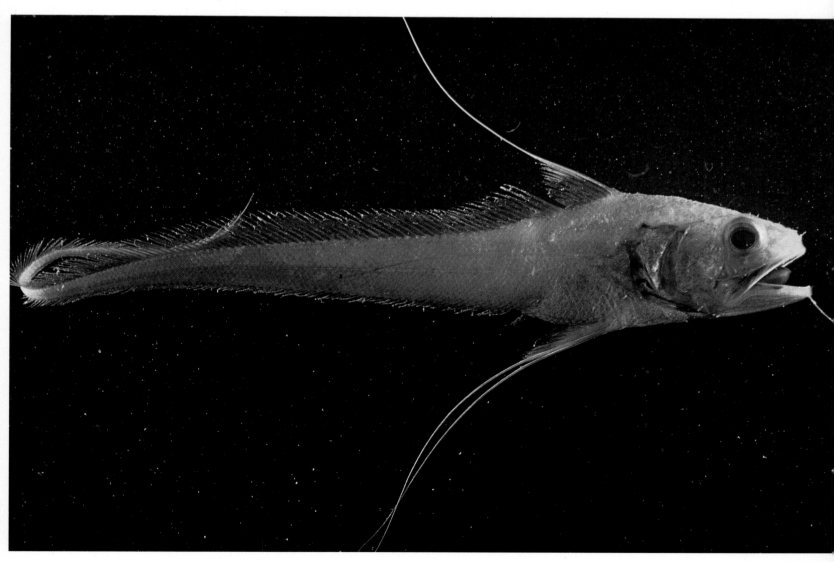

dependent on the food supply reaching them from the surface. It does so chiefly in the form of detritus. In the deepest regions of the ocean the benthos will be literally kilometres away from the surface and the animals themselves are completely isolated from it, being unable to swim and not having the necessary physiological systems to cope with the pressure changes even if they could do so. Life will, therefore, depend on certain types of animal being able to make use of the falling detritus. These, in turn, will be preyed upon by the various categories of carnivore. To collect the detrital particles some animals have delicate branching processes which gather up any particles falling on them. Sea-fans, deep sea corals and some of the echinoderms like the stalked sea-lilies may be able to do this. Others, like sponges, will pump water through their bodies and extract the food. Still others, such as the sea-cucumbers, will creep over the surface of the seabed and suck up the mud together with the organic matter it contains. There are great concentrations of bacteria in the benthic mud, or ooze as it is technically known, and they assist in

the break down of the detritus into its components which can be more readily digested by the bottom-feeders. Living among these particle and ooze feeders are the usual scavengers and carnivores. These will include crabs, prawns, and large sea-spiders. Some of these have long thin walking appendages. The echinothurid sea-urchins, which have flexible tests and appear quite squashy, probably also move around amid the more sedentary animals and ingest food resting on the mud.

Among the swimming organisms here one will expect to find various sharks and rays as well as bony fishes like eels and other special bottom fishes. Very few of these deep bottom dwellers have a larval stage that ascends to the surface waters to join the plankton. The pattern most commonly encountered is that which involves the development of large eggs rich in yolk from which juveniles probably hatch at a stage ready for bottom life. Probably the journey to the surface and back again at metamorphosis would be too full of risks. Many of the fishes undergo a developmental and juvenile stage alongside their parents. These reproductive strategies are just one

Above
This deep-water angler fish (*Melanocoetus*) shows the characteristic teeth set in enormous jaws. A modified fin ray forms the phosphorescent lure which is flicked about to entice prey within reach of the gaping mouth. The teeth are not so much for tearing at the prey but for preventing its escape from the mouth.

Above right
Sea-pens like this specimen of *Pennatula* are sometimes dredged from soft regions of the deep seabed. Many polyps are grouped on the branches, and a special region resembling a quill serves to anchor and support the colony.

Right
Lepidophanes guntheri is a lantern fish from deep water. A series of silvery spots mark the location of the photophores or light organs on its belly.

There is a great variety of deep sea-cucumbers. *Psycropotes depressa* shown here, has a strange extension on the upper side of its body which may act like a fin in currents and lift the animal temporarily from the bottom.

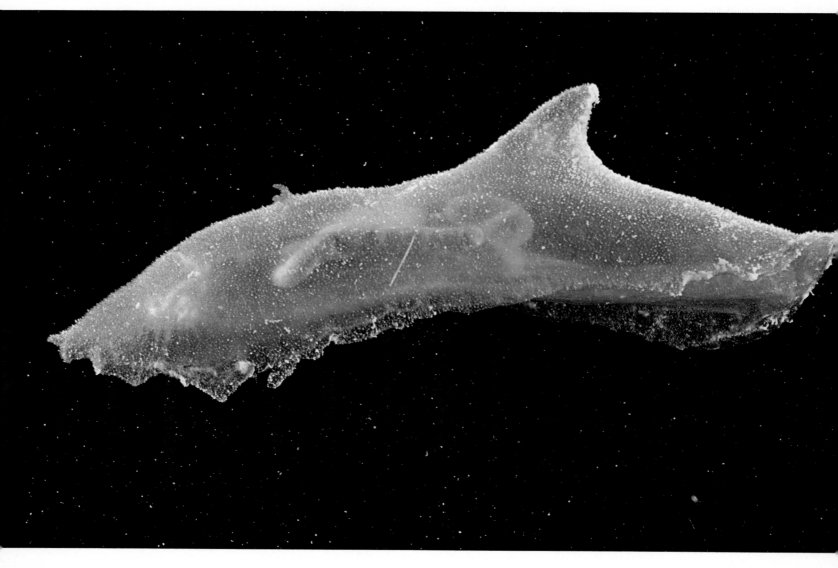

aspect of the many modifications that must be necessary for life on the ocean floor.

The depths of the ocean, more than any other region of the seas, emphasize the mystery and uncertainty that surrounds the lives of some marine animals. However, modern research techniques are improving man's knowledge of the oceans all the time. One of the most disturbing facts that such research highlights is the increasing extent to which the activities of man himself are affecting life in the sea in all forms. Pollution from industrial and urban areas on land, over-fishing, tourism and off-shore mining and drilling are all threatening the existence of many marine species. With the overall increase in the human population of the world's continents and the growing demand for raw materials and food, it is unlikely that these pressures on the sea will abate. Consequently, it is of the utmost importance that future plans for marine exploration take into account all damaging effects. It is only if a truly international approach along these lines can be achieved that serious pollution of the seas and oceans can be avoided.

Index